POLAR EXPEDITIONS

Polar Expeditions employs structural ritualization theory to show how rituals enriched the lives of crewmembers on 19 polar expeditions over a 100-year period. J. David Knottnerus identifies and compares failed, successful, and extremely successful missions in terms of participation in ritual practices and the social psychological health of crews, finding that social and personal rituals, such as work practices, religious activities, games, birthday parties, special dinners, or taking walks are extremely important in increasing crewmembers' ability to cope with the challenges they face, including extreme dangers, isolation, restricted environment, stress, lengthy journeys, and quite importantly the disruption of those practices that define our everyday lives. Besides contributing to our knowledge about polar expeditions, this research yields implications for our understanding of ritual dynamics in other situations such as disasters, refugee camps, nursing homes, traumatic experiences, and a new type of hazardous venture, space exploration.

J. David Knottnerus is Emeritus Regents Professor of Sociology at Oklahoma State University, Stillwater, Oklahoma. He has published extensively in the areas of ritual dynamics, social theory, social psychology, group processes, social structure, and social inequality. Much of his work in recent years has focused on the development of structural ritualization theory and research, which analyzes the role rituals play in social life.

"This book is an amazing exploration in its own right. The book brings together the rigorous Theory of Structural Ritualization and a careful and clever methodology, literary ethnography and content analysis. Nineteen expeditions are identified and analyzed and of those, nine are classified as failed, in the sense that the social fabric was irrevocably damaged, social rituals were rare, and conflicts, often violent, were common. Counter to these failed cases, in successful expeditions, collective rituals involving all crew members were common and, indeed, often planned by leadership. This book is a compelling analysis of how social rituals drive social life and group cohesion. It is rare (and so welcome!) for such an entertaining book to also be such a theoretically rich book."

Jane Sell, *Texas A & M University*

"*Polar Expeditions* reveals captivating details behind a century's worth of exploration into the most treacherous regions on earth. With groundbreaking sociological analysis, the book specifies activities leading to extreme levels of success in the face of hazardous challenges and prolonged periods of isolation. Professor Knottnerus breaks new ground by using structural ritualization theory to examine how crewmembers conquered disruption by reestablishing aspects of their lives back home. The book also focuses on goal attainment through the development of social bonds by way of work routines, collective events, and recreation. This book is essential reading for anyone wanting to know how to lead diverse groups of people. Readers will learn about open communication, developing grit, and generating solidarity, issues more timely now than ever."

Jason S. Ulsperger, *Arkansas Tech University*

"*Polar Expeditions* marks a watershed in social analysis. Knottnerus' book establishes ritual practice as the missing link in understanding how crews adapt and crewmembers cope with the extreme confinement and isolation imposed by hazardous ventures."

Jeltje Gordon-Lennox, *Psychotherapist and Editor of Coping Rituals in Fearful Times*

"The Polar regions are the harshest environments on earth and expeditions into these regions face extreme conditions, isolation, and other unique challenges. What makes some expeditions succeed when others fail? Professor J. David Knottnerus provides valuable insights into this question by examining the role of rituals in providing meaning and focus during such hazardous undertakings. His book, *Polar Expeditions: Discovering Rituals of Success Within Hazardous Ventures,* examines logbooks, diaries, and other first-person accounts from 19 expeditions occurring between 1850 and 1950 and applies Structural Ritualization Theory (SRT) to demonstrate the importance of rituals in shaping morale, comradery, and social cohesion among crewmembers. Expeditions that were able to develop new rituals were more successful than those who lacked robust reritualization practices. The result is a fascinating recounting of these expeditions as well as a clear advancement of SRT."

Duane Gill, *Virginia Tech University*

POLAR EXPEDITIONS

Discovering Rituals of Success Within Hazardous Ventures

J. David Knottnerus

Routledge
Taylor & Francis Group

NEW YORK AND LONDON

Cover image: © Shutterstock

First published 2023
by Routledge
605 Third Avenue, New York, NY 10158

and by Routledge
2 Park Square, Milton Park, Abingdon, Oxon, OX14 4RN

Routledge is an imprint of the Taylor & Francis Group, an informa business

© 2023 Taylor & Francis

Library of Congress Cataloging-in-Publication Data
A catalog record for this title has been requested

ISBN: 978-1-032-32290-2 (hbk)
ISBN: 978-1-032-32289-6 (pbk)
ISBN: 978-1-003-31431-8 (ebk)

DOI: 10.4324/b23044

Typeset in Bembo
by KnowledgeWorks Global Ltd.

To the members of all expeditions, past, present, and future

CONTENTS

3 Failed Expeditions   35

J. David Knottnerus, Kevin Johnson, and James D. Mason

4 Successful Expeditions   85

J. David Knottnerus, Kevin Johnson, and James D. Mason

ACKNOWLEDGMENTS

Years ago, I began to question how people cope with major changes in their lives such as taking part in an expedition where one's ordinary life and many of the ritualized behaviors a person normally participates in are disrupted. To better understand this topic, I oftentimes drew upon research others and I had conducted. I want to express my gratitude to all those I have worked with on projects dealing with ritual dynamics along with those scholars who have contributed to our understanding of ritual. In that regard, I should acknowledge the major contributions of two former doctoral students, Dr. Kevin Johnson and Dr. James Mason, who helped me investigate the role of rituals among crews on polar expeditions. And I would like to thank my wife, Frédérique Van de Poel-Knottnerus, Professor of French and French Literature, who has collaborated with me on various studies examining rituals in society and has always strongly encouraged my endeavors in this regard. I should point out that it has taken a number of years to read and analyze all the written accounts about the 19 polar expeditions examined in this study. For this reason, I want to express my gratitude to Dean Birkenkamp, editor, who has been extremely helpful, supportive, and patient as I prepared this book. Lastly, I would like to acknowledge and express my respect for all those persons who kept records about the polar expeditions they were members of. Without these oftentimes detailed accounts, sometimes written during periods of great uncertainty and danger, this study could never have been carried out.

1

INTRODUCTION

Expeditions and Rituals

In recent years, there has been a great deal of interest in polar expeditions, especially those in the nineteenth and early twentieth centuries. Witness the recent attention given by various news agencies to the discovery of some of the remains of the ill-fated Franklin expedition in which two ships were trapped in ice in 1846. For over a century and a half, people have speculated about the circumstances surrounding the disappearance of the two vessels in Canada's Arctic – the HMS Erebus and the HMS Terror. Both ships were part on an expedition led by British Captain Sir John Franklin to discover the Northwest Passage, a route that would provide a quick passage from the Atlantic to the Pacific oceans. Since their disappearance approximately 30 expeditions have tried to locate the boats with many more people losing their lives in the process. Finally on September 7, 2014, one of the two vessels was located using a remotely operated underwater vehicle. No doubt in the years ahead, attempts will be made to search the mostly intact shipwreck in hopes of finding books, records, and other items.

Another example of interest in expeditions concerns Ernest Shackleton and his journey to Antarctica (1912–1915). In recent years, several best-selling books as well as a television mini-series have focused on this mission. Actually in 2013, a group of explorers recreated Shackleton's famous Antarctic journey in which his ship, the *Endurance*, was crushed by ice and sank, resulting in a protracted journey over ice, oceans, and mountains to reach safety. Using the same route and a similar boat and equipment, this modern-day undertaking led by Tim Jarvis (2013) retraced Shackleton's voyage.

Of course, people's interest in polar expeditions can be seen in the numerous accounts of such journeys that have been published over the last century

DOI: 10.4324/b23044-1

or so. Indeed, the appeal of these books, especially first edition copies, is reflected in the market prices of these works, some of which sell for hundreds of dollars or more.

Contained in these written accounts are many fascinating details about the dangers faced by crews, the rugged environment, navigational and sailing information, and nautical decisions made by their leaders. But beyond all of that, the careful reader can also discover in the descriptions of some expeditions many discussions of celebrations, special dinners, and other activities, which at first may appear quite superfluous. Considering their need to pack several years' worth of food, clothing, building supplies, fuel, scientific gear, and other necessities, it might seem surprising that a number of these missions made room for, among other things, musical instruments, and luxuries such as wine, liquor, and chocolate. The Antarctic and Arctic regions are the harshest climates on earth, and given the severity of the situation, one might ask why did they bring such seemingly trivial items when they could have increased their margin of safety and brought more food or fuel instead?

I contend that the rituals associated with these items increased people's ability to cope with the extremely dangerous challenges they faced and the large-scale changes in their lives that ensued when they became members of an expedition. These rituals played an essential role in helping crewmembers to manage the life-changing and life-threatening conditions they encountered on their hazardous journeys. In many ways, this was due to the uplifting impact these activities had on individuals and the crews as a whole.

To better understand how this occurred, I employ a sociological/social psychological perspective – structural ritualization theory (SRT) – that focuses on the role ritual plays in social life (Knottnerus 1997; [2011] 2016). This perspective provides a framework for analyzing a large number of accounts of Arctic and Antarctic expeditions carried out in the late nineteenth and early twentieth centuries. By applying the theory to the evidence, we are able to shed light on what life was like for these crews both personally and socially.

The success or failure of these expeditions can then be assessed using broad categories reflecting social integration and the nature of the relations among crewmembers. The social attributes to be examined are emphasized by various social psychological and sociological traditions ranging from research on group processes and social interaction to classic social theorists such as Emile Durkheim.

In addressing these issues, abundant commentary and examples drawn from the sources examined by two former doctoral students (cited as coauthors in Chapters 3–5) and myself are presented describing the challenging conditions these expeditionary ventures faced, the interpersonal dynamics between people, the presence or the absence of rituals among crewmembers, and the effects of these practices on them.

More precisely, Arctic and Antarctic expeditions between the mid-nineteenth and mid-twentieth centuries often consisted of crews who were subjected to extreme hardships. These men and in certain cases women were forced to live in close quarters over extended periods of time as they fulfilled their sometimes dangerous duties. Thanks to the exploratory and scientific nature of the expeditions meticulous memoirs, diaries, and logbooks were often kept which detail what occurred during these journeys. These records provide an abundant source of evidence and fascinating insight into the crews' daily lives.

To conduct this study, we have analyzed the records of numerous expeditions using a qualitative method of content analysis. In this process, which literally involves reading line-by-line people's accounts of these journeys, records are searched for evidence of ritualized practices as suggested by SRT.

Guided by this approach, different kinds of rituals can be identified and described. These practices are then subjected to further analysis with regard to their social psychological impact on the members of the missions. A number of issues are examined, including how evidence shows the presence or the absence of rituals had a direct bearing on the success or failure of the crews and in some cases the expeditions as a whole. In the conclusion, the relevance of these findings and implications for further research are also explored, including possible applications not only to polar expeditions but other hazardous endeavors as well.

In sum, this investigation is grounded in a theoretical perspective dealing with ritual dynamics in which a great amount of material drawn from the descriptions of many expeditions are presented to corroborate the ideas discussed here.

Rituals

The array of ritual practices and the degree to which crews performed them on certain polar expeditions is truly striking. A few examples can be presented right now of these activities.

One type of ritualized behavior involves repeated scientific practices. This was the case even for troubled missions such as the Jeannette Expedition of 1880 to be examined later. After becoming lodged in polar ice, the ship began to leak past the point where bilge pumps could keep up. This was a dire emergency as the ship was close to sinking thousands of miles from any port. Additional pumps were hastily constructed; however, and soon afterwards they resumed normal scientific (including geographical) observations, a practice the captain zealously adhered to. As we shall see a devotion to such work rituals even in the face of deadly threats is evident in a number of expeditions.

A very different type of activity involved celebrations in which crewmembers stopped their labors, came together as a group, and enjoyed themselves. On particular missions no expenses were spared, e.g., all happily shared fine foods and drink that had been saved for such events.

Religious rituals can also be found on some expeditions. In this regard, Bocock (1970: 288) suggests religious rituals bring people together and increase the awareness that they are members of a group: "In many of these rituals, people are made aware of 'society,' its laws and the obligations of its members, of something more powerful than themselves standing over and above them." Given the isolation and dangers these explorers experienced, it is easy to appreciate how these activities might appeal to some. Here I would note that military procedures sometimes mandated or encouraged religious services being held. This was more overtly noticeable in the earlier expeditions lessening to some degree in later accounts. Either way such rituals can be extremely moving for participants. Of course, this was a fundamental insight of Emile Durkheim ([1915] 1965), who explored the nature of religion, collective celebrations, and ritual a century ago.

It should also be appreciated that some expeditions included both military personnel and scientists (and other groups such as civilian seamen). This created a situation where parts of the crew were naval officers and seamen, subject to military law, while the scientists had a less formally recognized hierarchy. Still the rituals associated with both groups could influence the behaviors of individuals and even others who were not part of the crew. For instance, in the disastrous Jeanette expedition to be later examined, the dead military crewmen that had been haphazardly buried by the starving survivors were later dug up by rescuers and reburied in a more appropriate spot. In accord with accepted burial rituals, the bodies were ordered in the grave according to their rank.

Lastly, in a different vein, projects were sometimes initiated that helped prevent inactivity during the long polar nights. Some crews created newspapers with amusing stories and discussions about holidays. Midwinters and Midsummer's Day were especially important, as they marked the high or low point of the seasons, depending on the location of the missions. Another holiday, which was usually mourned and then celebrated, was the disappearance and reappearance of the sun.[1]

These and many other rituals will be much more extensively examined in subsequent chapters, especially in respect to how successfully or unsuccessfully crews managed the challenges facing them. By doing so, we will be able to more clearly determine whether the presence or the absence of such practices had a direct bearing on the crew's performance and state of mind.

Research on Polar Expeditions and Other Hazardous Ventures

Before going any further we should appreciate that research has been conducted through the years about polar expedition teams and other similar groups. Much of this important research, however, has emphasized the psychological and personality attributes of people and how these traits influence the behaviors of crewmembers as opposed to the ritual dynamics of crews and how these rituals influenced crewmembers. These studies examine an array of psychological factors ranging from achievement motivation and well-being to cognitive changes, anxiety, mood, emotional concerns, states of consciousness, stress, and depression. Employing various methodologies, they have studied different polar missions composed of crewmembers from different cultures. And while oftentimes the crews studied involve men, some research has focused on all women crews or mixed gender groups (Leon, McNally, and Ben-Borath 1989; Leon 1991; Mocellin and Suedfeld 1991; Kahn and Leon 1994; Peri, Scarlata, and Barbarito 2000; Weiss, Suedfeld, Steel, and Tanaka 2000; Palinkas 2003; Palinkas and Suedfeld 2008).

Other related research has focused on human behavior in different environments such as space or ocean depths marked by isolation, confinement, and danger. These studies are sometimes informed by investigations of expeditions. This research ranges from work that is oftentimes psychological in nature to a limited number of studies involving sociological analyses (e.g., Emerson 1966; Suedfeld and Steel 2000; Suedfeld and Weiss 2000; Dion 2004). Again we find that while valuable, this research does not direct attention to the ritualized practices of persons and how these activities influence them.

Furthermore, there are a limited number of studies found in specialization areas other than psychology or sociology. One example would be a content analysis of diaries and correspondence of Arctic whalers and explorers written in the nineteenth and early twentieth centuries that focuses on representations of family and social origins. Johnson and Suedfeld (2012) found that these individuals often engaged in activities similar to those prevalent back home. While this investigation does not analyze such activities in terms of rituals it is suggestive in that it indicates that people will often try to recreate taken-for-granted practices during hazardous ventures.

Finally, I should make special note of Jack Stuster's *Bold Endeavors: Lessons from Polar and Space Exploration* (1996). This work provides some very interesting descriptions and quite perceptive discussions of various polar expeditions while presenting a number of recommendations for space travel. At the same time, the book does not focus on ritual dynamics, is not formally grounded in a sociological or social psychological perspective and does not examine in a methodologically systematic manner the practices of expeditionary explorers. In effect it does not distinguish between expeditions in terms of whether and the degree to which crews engaged in ritual practices. All of these points stand in marked contrast to the approach taken here.

Conclusion

In sum, this project is unique because no other book or other kinds of published research that I am aware of provide a sociological/social psychological or more broadly speaking social scientific analysis of expeditions and the ritual dynamics of crews.

To accomplish this goal, I will discuss in the next chapter the assumptions and theoretical framework of the approach used to understand the social dynamics of expeditions and the research strategy used to conduct this study. The subsequent three chapters will then examine what life was like on a large number of polar missions. These chapters will focus on the different kinds of rituals crews may have engaged in and their importance. These sections will be arranged in terms of how unsuccessful or successful expeditions were in terms of the social dynamics of crewmembers. Lastly, in the conclusion, the findings of this investigation will be summarized, and implications addressed including directions for possible future research. And I will discuss the relevance of this investigation for other types of hazardous ventures involving extremely dangerous conditions, great amounts of stress, high degrees of isolation, and restricted environments in which people work, eat, sleep, and interact with each other.

Note

1 Depending on the latitude in the North, the sun will disappear for several months in the winter, then re-appear as the days steadily get longer until the sun stays out for several months at a time. Eventually, the sun begins to disappear again, and the days get shorter until it disappears for several months. The opposite occurs in the South. This pattern of sun and no sun can negatively impact the emotional and physical well-being of some people and lead to complaints that the constant light interferes with their sleep patterns. Very likely the lack of light led different expeditions to report cases of individuals suffering from what later became known as arctic sickness or seasonal affective disorder. This condition is caused by the lack of sunlight on the retina, which then signals changes in brain chemistry and causes physical and psychological symptoms. One idea worth considering is whether different kinds of ritual practices might help alleviate certain of the symptoms that this disorder could contribute to.

2

RITUAL THEORY AND METHODOLOGY

Life on Polar Expeditions

Since this investigation rests upon the assumption that rituals play a major role in social life, we must first discuss how they are significant, their place in sociological and social psychological explanations, and the particular perspective that provides the framework for this study.

The idea that rituals are crucial to human behavior is consistent with the arguments of different scholars who have used this idea to study society such as Warner (1959; 1962), Durkheim ([1915] 1965), Goffman (1967), Hocart (1968), Douglas (1970), Kertzer (1988), Bell (1992), and Collins (2004). However, despite the contributions of social scientists such as these, the importance of ritual has been oftentimes under-appreciated in sociology and other disciplines.

This is largely due to the conventional interpretation of rituals in sociology, the social sciences, and contemporary society. For example, it is commonly believed that rituals are found mostly in premodern societies, not modern societies, which is due in part to social evolutionary suppositions which assume the latter is much more rational, advanced, and superior, rituals are unchanging or static in nature, rituals occur only in religious or sacred settings, and rituals are of minor importance because they are the product of more important social influences or are simply not involved with these more significant dimensions of society such as the political arena or the economy. These assumptions, especially the last one, lead to the conclusion that rituals have little effect on people, interpersonal relations, social structure, institutions, the exercise of power, and collective events, especially in contemporary society.

For these reasons rituals are often thought to have limited explanatory value and are downplayed if not simply ignored in our analyses of social life. In other words, they remain in various ways unseen and off the proverbial radar of those who seek to explain how societies work and groups operate.

DOI: 10.4324/b23044-2

Structural ritualization theory (SRT), on the other hand, emphasizes the role rituals play in a society (Knottnerus 1997; 2005; 2009; 2010; 2014 a, b; [2011] 2016). It assumes that daily life is characterized by an array of personal and social rituals. These activities, which we often take for granted and do not think about or question, define in fundamental ways the nature of our lives both in terms of who we are and how we relate to others. Indeed, these rituals help stabilize social life while expressing symbolic meanings which give significance to our actions. Thinkers such as Emile Durkheim ([1915] 1965), Peter Berger and Luckmann (1966), Erving Goffman (1967), Anthony Giddens (1984), and Randall Collins (2004) have through the years made similar arguments about the centrality of rituals or routines in human behavior. SRT builds on their contributions.

Consistent with these arguments SRT stresses how ritual, whether occurring in small groups or organizations, can lead to consequences unanticipated by group members, while both being fed by and feeding into other parts of society including larger groups and societal arrangements. Consequently, this perspective is directed to rituals which occur in various social settings, e.g., face-to-face interaction (or interpersonal relations), small groups, organizations, society as a whole, and even globally.

Ritual as a Missing Link

More precisely SRT is grounded upon a very different set of assumptions from those that underlie much of the work in sociology and other related disciplines. Basically I assume that ritual provides what I have referred to elsewhere (Knottnerus [2011] 2016) as a "missing link" in sociological thought and research.

While not all of these presuppositions need to be discussed, I should mention several of them. Succinctly stated these assumptions include the ideas that rituals are found in both premodern and modern societies, rituals occur in both secular and sacred or religious contexts, rituals are dynamic in nature and subject to change, rituals can be of profound significance in social life, and rituals have great explanatory value.

SRT also assumes that the idea of ritual can provide a common analytical framework and language to study the dynamics of different groups. In other words, it can help us better understand in a coherent and systematic manner the workings of society. As such ritual is a concept that has potential explanatory relevance for the many-sided nature of social reality. Stated somewhat differently SRT uses an analytical approach, which employs the concept of ritual to explain many kinds of activities. It provides abstract formulations, which address ritual dynamics occurring in various settings such as youth groups in schools, health care facilities, corporations, sporting events and organized recreational activities, ethnic and racial communities, slave societies within plantations, political or religious groups, and social movements including those engaged in genocide.

Furthermore, SRT uses multiple research strategies. Research employs different approaches involving both qualitative and quantitative forms of evidence such as historical-comparative analyses, case studies (contemporary and historical), interviews, content analysis of primary sources, field research/ethnographies, experimental research, surveys, and literary sources and accounts including novels, autobiographies, biographies, memoirs, travelogues, and diaries. It is grounded in an appreciation for the strengths of different methods, an awareness of how varied strategies can complement each other, and the significance of examining cases in different locations.

In conclusion, this perspective is of value because it offers a set of ideas dealing with a key component of behavior that can be used to examine many dimensions of social reality. It provides abstract concepts that can be applied to different groups and situations in a wide range of studies. This is critically important given the enormous complexity of human behavior and what some (Phillips 2001; 2009) consider our less than complete success in addressing that complexity because of the specialized approach to research often taken in the social sciences.

Development of SRT

SRT has been in development for a number of years and has expanded in scope as it has progressively examined a greater number of issues and groups. I will begin with a very brief description of the first research and some of its core concepts.

The first formulation of the theory (Knottnerus 1997) emphasized how rituals rest upon cognitive schemas and express symbolic meanings or themes. I formally defined ritualization and presented a set of factors that influence the importance of rituals in a social setting and which explains how social action and social structure are reproduced or transformed. In that regard, several elements of the original theory should be defined. An *action repertoire* is a set of socially standardized (regularly engaged in) practices. A *schema* is a cognitive structure or framework. Using these terms, a *ritualized symbolic practice* is an action repertoire that is schema driven. Ritualized symbolic practices (RSPs) involve common social behaviors where people engage in regularized and repetitive actions oftentimes when interacting with others. RSPs are ubiquitous in nearly all forms of social life. SRT argues that these ritualized action repertoires rest on cognitive schemas, which may not be immediately aware to the person, yet still communicate various thematic meanings.

The original theory emphasizes the importance of embedded groups or groups that are nested within a more encompassing collectivity, e.g., an informal youth group in a school, a problem-solving group within a formal organization such as a corporation or government office, or a slave society within a plantation. The theory focuses on the ritualized practices performed in a wider social environment that acquire significance for actors and then become part of individuals' cognitive scripts for acting in their immediate, embedded social world. In this way, ritualized practices develop in ways that may be similar to and confirm the patterns of behavior in the wider social environment, i.e., are reproduced.

Counter to what one might intuitively expect, different research findings show that reproduction occurs in embedded groups even when no incentives exist for doing so, individuals in such groups are briefly exposed to RSPs in the wider environment, only some members of the nested group are exposed to those ritualized activities, group members verbally express their opposition to the wider social milieu, and it does not serve the interests of group members. These findings have been obtained in studies of an array of settings involving students in schools, experimental groups, slave societies, and youth in ancient Sparta.

Based upon the original theory SRT has evolved in different directions examining ritual dynamics in contemporary societies and in the past. The upshot is that several lines of theory/research development are under way each of which builds upon some of the original formulations dealing with structural reproduction. And a number of studies have been carried out providing tests, exemplifications, and applications of these extensions. All of this work is committed to the development of abstract concepts, grounding these ideas in empirical evidence, and using this knowledge to alleviate social problems.

This research basically falls into eight categories. These areas of study examine different aspects of ritual dynamics and both the positive and negative effects of ritualized practices.

As just discussed, research into the reproduction of RSPs and social structure within groups constitutes one area of study. Studies of this issue range from investigations of male and female youth societies in nineteenth century, male and female French elite educational institutions (Knottnerus and Van de Poel-Knottnerus 1999; Van De Poel-Knottnerus and Knottnerus 2002), slave societies in American slave plantations, (Knottnerus 1999; Knottnerus, Monk, and Jones 1999) and the cultivation of extreme militaristic behaviors among youth in ancient Sparta (Knottnerus and Berry 2002) to experimental task groups (Sell, Knottnerus, Ellison, and Mundt 2000).

A different body of research focuses on disruption, deritualization, and reritualization. Here the focus is on breakdowns of social and personal rituals, their consequences, and the ways people may cope with such experiences by reconstituting old or new ritualized practices. Studies dealing with this issue have dealt with internment in concentrations camps (Knottnerus 2002; Van de Poel-Knottnerus and Knottnerus [2011] 2016), disasters (Thornburg, Knottnerus, and Webb 2007; 2008; Bhandari, Okada, and Knottnerus 2011), a laboratory experiment involving disruptions in task groups (Sell, Knottnerus, and Adcock-Azbill 2013), dark ages/periods of ecological degradation (Sarabia and Knottnerus 2009), the displacement of youth during China's Cultural Revolution (Wu and Knottnerus 2005; 2007), a research review (Knottnerus 2005), how the failure to enact rituals can contribute to mass homicide (Ulsperger, Knottnerus, and Ulsperger 2017), and a study of ritual and social control by the Khmer Rouge in Cambodia (Delano and Knottnerus 2018).

Another line of work has begun to address the role of emotions in rituals and collective episodes. A theoretical model has been presented of emotional intensity, group commitment, and solidarity in collective events, (Knottnerus 2010) along with a discussion of collective emotions in religious and non-religious groups (Knottnerus 2014b; see also Meij, Probstfield, Simpson, and Knottnerus 2013). And several new concepts have been presented concerning collective pride, emotions, and ritual (Knottnerus 2014a).

Other studies have dealt with how ritual can influence the development of identity in different groups, social inequalities, and the experiences of immigrants. This research focuses on the maintenance of traditional female identity in Malawi (Minton and Knottnerus 2008), ethnic identity and a bi-ritual character among some Chinese Americans (Guan and Knottnerus 1999), and multi-ritual identity and first generation Asian Indian Americans (Sen and Knottnerus 2016).

In a different vein a significant amount of attention has been given to the enactment of ritualized practices in formal organizations oftentimes involving organizational deviance. The ritual dynamics of two organizations in particular have been studied: ritualized maltreatment and neglect in nursing homes (Ulsperger and Knottnerus 2007; 2008; 2009a; 2009b; 2011; 2013; 2021; Ulsperger, Knottnerus, and Ulsperger 2014) and financial and managerial deviance in the Enron corporation (Knottnerus, Ulsperger, Cummins, and Osteen 2006; Ulsperger and Knottnerus 2010). A model of Drug/DUI courts and ritual dynamics has also been presented (Liang, Knottnerus, and Long 2016) in addition to the study of mass homicide (Ulsperger, Knottnerus, and Ulsperger 2017).

SRT has also been concerned with strategic ritualization and the ways power and ritual can influence each other. Four ways individuals and groups may use rituals to achieve different goals have been identified: ritual legitimators, sponsors, entrepreneurs, and enforcers. Investigations of this topic focus on the strategic use of rituals in Italian-American ethnic communities (Knottnerus and LoConto 2003), ritual enforcement and power in Nazi Germany, the Orange Order, and Native American Pow Wows (Knottnerus, Van Delinder, and Edwards [2011] 2016), the Orange Order and parading (Edwards and Knottnerus 2007; 2010), a Chinatown community protest movement (Guan and Knottnerus 2006), the Notting Hill Carnival in London (Edwards and Knottnerus 2011), and techniques of control through the use of rituals by the Khmer Rouge (Delano and Knottnerus 2018).

Additional research has dealt with ritual dynamics that produce social inequality, social distinctions, and different forms of exclusion. This research has examined traditional ritualized behaviors of women volunteers in NGOs concerned with women's rights in India (Mitra and Knottnerus 2008), royal women in ancient patriarchal India (Mitra and Knottnerus 2004), golf, civility, class, and exclusion in America (Varner and Knottnerus 2002; 2010), and gender inequality in Malawi (Minton and Knottnerus 2008).

Finally, attention has been given to applied research, social policy, and interventions involving changes in rituals. Much of this research focuses on alterations in RSPs among staff and residents in nursing homes and a proposal for how to do this entitled the CARE model (Ulsperger and Knottnerus 2007; 2008; 2009a; 2009b; 2011; 2013; 2021; Ulsperger, Knottnerus, and Ulsperger 2014), and research that seeks to facilitate ethnic entrepreneurship among immigrants in Canadian society (Lin, Guan, and Knottnerus 2011). Another major concern deals with the ways rituals might help people cope with the disruptions (as previously discussed) that befall them in their everyday lives, disturbances that can lead to a host of troubles as in the case of disasters such as tornadoes or terrorist attacks (Thornburg et al 2007; 2008; Bhandari et al 2011; Johnson, Knottnerus, and Gill n.d.).

Besides these eight lines of research, attention is also being given to the ways rituals can promote persecution and impact the social dynamics of groups who persecute others (Knottnerus n.d.).

In sum, this research rests upon the assumption that rituals make up a key dimension of social behavior just as different perspectives focus on other aspects of social action, such as rationality focused on by social exchange theory and rational choice theory, symbolic interpretation emphasized by symbolic interaction, emotions addressed by the sociology of emotions, and impression management, i.e., dramaturgical presentations of self, stressed by Erving Goffman. Succinctly stated the approach used here assumes that *ritual is like an engine, which drives much social life*, sometimes quite forcefully (Knottnerus [2011] 2016).

In examining this topic SRT focuses on three interrelated goals: (a) the development of theoretical ideas, (b) empirical research which corroborates and exemplifies these principles, and (c) using this knowledge in an applied manner to alleviate social problems, inequality, and the dehumanization of people.

Expeditions and SRT

As important as all of these issues are, there is a particular kind of occurrence in many people's lives that also deserves attention. This type of situation involves long term, stressful endeavors characterized by extreme isolation and confinement. Such experiences present people with an extremely unique set of circumstances, which can be strange, unsettling, and particularly difficult to deal with. These kinds of situations, which can last months if not years, are also the concern of SRT.

And this is where our interest in expeditions comes to the forefront. Expeditions present us with a vivid and telling example of a hazardous venture where people are engaged in an enduring, highly stressful enterprise marked by extreme confinement and isolation. It is a situation that by its very nature is extremely different from the conditions we normally encounter in our daily lives; it is a situation which calls upon individuals to do things and act in ways that they usually don't. Expeditions involve a radical departure from our taken-for-granted normal lives. And it is a situation in which rituals may help, perhaps greatly help, people cope with such circumstances.

In suggesting this is so, the present study rests upon the basic assumption of SRT that rituals are a crucial part of our lives. As already stressed this approach assumes that our everyday life is defined by an assortment of personal and social rituals. These practices help produce stability in social action while communicating symbolic meanings, which give significance to our behavior. By investigating expeditions, we examine one way the daily RSPs people participate can be disrupted and how people may cope with such occurrences.

In making this argument, I suggest that the disruption of patterned, ritualized behaviors is a problematic situation for both individuals and groups. This is when the importance of constructing or reinstituting rituals is most evident. When individuals and groups can recreate new or old RSPs after such disruptions, this enables them to reconstruct their normal, daily lives. These actions establish newly organized personal and social patterns grounded in ritualized practices and relationships. Through the development of a new pattern of rituals people gain a meaningful understanding of their world, a focus or sense of direction, and an overall stability in their everyday lives. And, when people do so these rituals not only provide meaning, coherence, and direction to our behavior, they also divert our attention away from conditions that may be beyond our understanding, chaotic, dangerous, miserable, stressful, disgusting, boring, or just uninteresting.

In other words, groups and individuals who can construct new or reestablish old, ritualized practices after a disruption are best able to *adapt* to and *cope* with the situation, they are in. These group and personal rituals act as a *buffer* or *cushion* to disruptive events.

Disruption, Deritualization, and Reritualization

More precisely SRT focuses on this topic by addressing what are referred to as disruption, deritualization, and reritualization. *Disruption* concerns occurrences, e.g., events or developments, which interrupt or disturb the RSPs that groups and individuals engage in. Many types of occurrences can result in the breakdown of rituals, the relations between individuals and groups, and social arrangements, large and small. These events can differ in their strength and may deeply influence the daily lives of actors. This situation can then lead to deritualization. *Deritualization* refers to the breakdown or loss of previously engaged RSPs for both individuals and groups. It involves the collapse or cessation of rituals that occur in our normal everyday lives.

I would stress that RSPs involve at least two key dimensions: action and symbolic or cognitive frameworks, which possess meanings (these two dimensions reflect the original definition of RSPs as schema-driven action repertoires). These are central components of RSPs, which are subject to breakdown or loss during deritualization. Deritualization affects both behavior and meaning in rituals. It results in the collapse and cessation of both symbolic meaning and action in people's ordinary lives. Certainly, while both are extremely important and usually involved in deritualization the degree to which they are present can vary. So, deritualization may reflect to a greater extent the loss of meaning or the breakdown of action. Or as is most likely the case, it may involve both to a similar degree.

A critically important way people can cope with disruption and deritualization is to re-engage in rituals. Ritual enactments or performances act as buffers that allow people to handle the harmful and threatening consequences of deritualization. They help cushion or shield individuals from this negative situation. *Reritualization* involves the re-creation of RSPs following disruption and deritualization. By reconstituting rituals following disruption and deritualization, human beings can restore crucial action repertoires of a social and personal nature. They allow people to manage the situation they are in.

Reritualization, therefore, makes it possible for individuals and groups to re-establish meaning and a direction, or focus on their behaviors. These renewed rituals, whether involving totally new or previously engaged in RSPs, restore coherence in our actions and cognitions. They produce clarity in people's thinking and designs for action. This re-grounding in an orderly and familiar taken-for-granted world, recreates an essential sense of reassurance, security, and stability.

While disruption, deritualization, and reritualization are varied in nature they are also quite common. In looking at how all three are found in polar expeditions, we gain a greater understanding of how they occur not only in such ventures but in the many other situations humans may find themselves in that possess features similar to the hazardous missions we examine here.

Model of Ritual Strength

In better understanding how all three operate a key issue concerns how rituals recreated after disruption and deritualization may differ, how and why they differ, and the extent to which they impact the crews of expeditions. SRT provides us with a model that allows us to assess how rituals may vary and their impact on people.

This approach formally identifies four elements – salience, repetitiveness, homologousness, and resources – that influence the dominance of RSPs in a social setting. These factors affect the strength of rituals in groups and other social milieus and their effect on persons (see Knottnerus 1997; [2011] 2016:20–24).

Salience is defined as the degree to which an RSP is perceived to be central to an act, action sequence, or bundle of interrelated acts. As already discussed, RSPs are grounded in symbolic schemas (models). Symbolic meanings or themes are communicated or expressed through the ritualized act. Thus, salience refers to the degree to which an RSP within a bounded social arena or social setting, i.e., "domain of interaction," is conspicuous, noticeable, or prominent. In other words, ritualized behaviors can differ in the extent to which they are salient or visible.

For instance, if the interaction among leaders and crewmembers in an expedition prominently consists of strict discipline, an authoritarian demeanor, great social distance, and is of an extremely demanding nature, this would be a very salient kind of ritualized behavior, which clearly emphasizes the importance of social power and the hierarchical nature of relations. These highly salient RSPs and the symbolic meanings expressed through them involving power, authoritarian relations, and inequalities among the crew are clearly recognized and understood by all.

Repetitiveness refers to the relative frequency with which an RSP is performed. This part of the theory stresses how ritualized practices can differ in the degree to which they are repeated. The occurrence of an RSP can range from rarely or never occurring in a social arena to its being engaged in numerous times. For example, crewmembers in an expedition might repeatedly engage in behaviors that are more personal in nature, informal, and friendly dozens of times a day, while another crew might only engage in ritualized behaviors that are of a more authoritarian and impersonal quality. Or crewmembers might sponsor different kinds of games, contests, and recreational events on a weekly basis while a different crew might rarely join together in such collective activities.

The degree to which rituals are repeated may at first glance seem to be a very simple attribute and not have great significance. But the extent to which people perform RSPs can be extremely important. The degree to which certain rituals are or are not engaged in can have a pronounced effect on social interaction, the types of relations that form in groups, and the overall state of mind or morale of individuals.

Homologousness involves the degree of perceived similarity among different RSPs within a social setting. Ritualized practices may exhibit to different degrees a perceived correspondence in their form and meaning. The greater their correspondence the more likely they will have the same impact or outcome. Stated somewhat differently these practices reinforce each other, enhancing their dominance in the setting and their effect on people.

For instance, an expedition crew could engage in various RSPs that are quite comparable in their meaning and form. Crewmembers might engage in ritualized practices such as celebrating each other's birthday, preparing plays or shows for amusement, and playing games together. While these are different RSPs they are similar to each other in that they emphasize the group as a whole, crewmembers working together with each other, and individuals enjoying each other's company. In contrast another crew might regularly eat their meals together, but not engage in any other ritualized activities such as playing together or collectively working in different projects apart from work. Such a situation might be due to rigid distinctions between officers and the rest of the crew, scientists and military personnel not socializing with each other, or differences between individuals who are of different nationalities and cultural backgrounds. The first crew would, therefore, have a higher degree of homologousness than the second crew because of the greater number of similar rituals they participate in.

RSP *resources* refer to the materials needed to engage in RSPs that are available to actors. The more the resources are available to actors, the easier or more likely that individuals will engage in rituals. This factor stresses a fundamental point that rituals require an array of resources to be carried out. While ritual practices include symbolic meanings (or cognitions), they are at the same time achievements by human beings who need many things to act whether in groups or alone.

Resources comprise many different attributes, items, and commodities. They can include knowledge, intellectual abilities, interpersonal skills, physical strength, dexterity, organizational supplies, clothing, banners, flags, musical instruments, communication technology, and so forth. At the same time people differ in their ability to perform rituals and to be influenced by them. Some RSPs may require resources such as physical abilities, mental skills, or behavioral/interpersonal skills that individuals may not have, e.g., a young child who doesn't yet have the capacity to engage in a team/group sport or understand how to perform a sophisticated dance routine or musical production. People only take part in ritual practices that they can grasp, manage, and at the very least adequately perform.

To capture the variety of resources that may be used in ritual practices I distinguish between two types. "Human resources" are defined as the abilities and characteristics of actors perceived by group members to be of value (or to have utility) for him or herself or the group. Such resources could include features or traits such as reading ability, oral skills, musical talent, physical strength, or expertise in some area. "Non-human resources" are defined as all that is not human that is perceived by group members to be of value (or to have utility) for themselves or the group. Examples of this kind of resource could range from food, liquor, banners, writing materials, costumes, physical props, or different kinds of equipment such as scientific instruments or weapons for hunting to books and diaries. As we shall see nearly all of the resources mentioned and others have been used to varying degrees of success by expedition crews to create personal and social rituals.

These four factors influence the rank of RSPs in a social setting. *Rank* refers to the relative standing of an RSP in terms of its dominance. SRT argues that ritualized activities can vary in their strength or importance in any social milieu. Rituals can differ in their standing or dominance in a social setting whether that involves a group of adults working together on an office task, students interacting in a study group, or expedition crews engaging in some activity such as playing cards, caring for animals they brought with them, or sailing the ship they are on. These RSPs, the theory further argues, can be assigned a value, which signifies their degree of standing or influence. Rituals can, for instance, have a low, medium, or high rank.

To reiterate, SRT assumes that the rank of RSPs in a social setting or social environment is the sum of salience, repetitiveness, homologousness, and resources. The greater these four factors, the higher the rank of rituals in that setting which individuals are exposed to and/or engage in. The main point is that the higher the rank of RSPs in terms of these four elements, the greater the dominance or strength of these practices.

So as the rank of rituals increase, the greater their effect on people.

Stated in a slightly different way, the higher the rank of rituals in a social situation, the greater their influence on the thoughts and actions of individuals who are engaged in or exposed to these practices. The more highly ranked these activities, the more they affect people's cognitions and behaviors, and the dynamics of the groups they are a part of.

For this reason, when crewmembers engage in what we have previously referred to as reritualization, the higher the rank of these RSPs the greater the impact these rituals have on individuals. Rituals that are of high strength and are of a salutary or constructive nature will have a positive effect on crewmembers better enabling them to cope with the disruption and deritualization experienced during expeditions. These practices help buffer crews from the loss or a breakdown of previously engaged rituals and thus enable them to adapt to their very unique and hazardous situation marked by long-term isolation, confinement, danger, and stress.

In our examination of the accounts of different expeditions we will assess whether rituals occur and if so their rank. To determine the importance of rituals for crewmembers our primary focus will usually be on the first two factors – salience and repetitiveness. The analysis of these factors will be further enhanced through an examination of available evidence dealing with the resources that support these activities and the similarity or lack of similarity of practices, i.e., homologousness.

The Nature of Rituals

We will also direct our attention to the nature of the rituals crews engage in. This will be done in several ways. First, we will determine if reritualization involves reconstituted rituals or new rituals. In other words, when people engage in ritual activities, they may reconstitute rituals that occurred in their ordinary lives prior to the expeditions they are involved in. On the other hand, crewmembers may construct new, unique rituals. Both types can help individuals adapt to the disruptive, stressful nature of an expedition. Whether this is the case and the degree to which this may occur are questions we will address.

We will also identify the kinds of ritual practices crewmembers engage in. Here we will determine if the specific RSPs individuals participate in, fall into different categories such as religious rituals or music rituals. In other words, are there substantive differences in ritualized practices, i.e., ritual contents, and if so, what are they? In identifying the particular nature of practices various questions arise such as, are certain types of rituals more or less common and are certain kinds of rituals more important than others for crewmembers.

Additionally, attention will be given to the social and non-social nature of rituals (Knottnerus 2002). RSPs may be practiced collectively and exhibit an organized, sometimes formally organized, character. On the other hand, social rituals may be conducted in a less regulated, less organized informal manner. Moreover, individuals may engage in ritualized activities alone. These distinctions are categorized as formal or quasi-formal organized rituals, informal interaction rituals, and personal/private rituals.

Program of Ritualized Symbolic Practices

There are a number of recently developed concepts that can further help us understand the social dynamics of expeditions.

Especially relevant is the issue of whether and how leaders may contribute to the development of ritual activities. Actually, it is possible that a leader may cultivate not just one or two rituals among the crew but a wide range of such behaviors. A relatively new concept – program of ritualized symbolic practices – addresses this concern because it directs our attention to how individuals and groups may strategically or tactically employ rituals to achieve different goals and express key ideas.

A *program of ritualized symbolic practices* is defined as a collection of RSPs strategically used by a group or individuals such as group leaders to achieve certain objectives (for a discussion of this concept and its application to various groups including the Nazi party, the Orange Order, and Native American Powwows see Knottnerus, Van Delinder, and Edwards ([2011] 2016). This

concept emphasizes that individuals, groups, or organizations may have differing abilities to employ an assortment of rituals to achieve one or more ends,
such as enhancing the power or collective identity of a group or increasing an
organization's visibility and prestige. Of course, another possibility is that the
leader of an expedition may foster rituals in the effort to improve the morale
of crewmembers and their ability to work together.

RSPs that comprise a program can, however, differ in terms of their type,
number, and frequency they are engaged in. Practices within a ritual program
may also complement or be similar to one another in regard to the symbolic
themes expressed through these acts. More generally ritual programs can differ in their level of complexity. For example, a ritual program could involve
many dominant RSPs, which complement each other. In contrast another
program might be far simpler containing only a few practices. That is why
ritual programs can be usually identified along a spectrum with those at one
end of the continuum having the greatest complexity, e.g., the most kinds of
RSPs, highest frequency of performances, greatest salience, and so on, and
others at the opposite end of the spectrum being the least complex.

Furthermore, great differences can exist among groups and individuals,
which strategically promote ritual programs. For instance, variations may exist
in the kinds of groups implementing RSPs, the resources they possesses to
enact rituals, the extent of organization and coordination that allows them
to create rituals, the degree to which practices are institutionalized, formally
assigned, and carried out by group members, and the degree to which RSPs
are due to the actions of a single person or highly controlled group or are the
result of different individuals and subgroups that are working together, nego-
tiating and compromising with each other, or in disagreement. These differences deal with the capabilities and organizational structure of individuals and
groups that promote ritual programs.

It is quite possible that expeditions differ in regard to both the complexity
of ritual programs and how they are encouraged or enforced. One might
find, for instance, that in some cases few if any rituals occur, while in other
cases crews engage in a large number and range of RSPs. So, too it may be
that in certain cases leaders or other crewmembers give little or no attention to organizational conditions, which would make it possible to engage
in rituals. Conversely, we might find that in other missions much greater
consideration is given to the planning of rituals, the stocking of resources to
conduct these activities, structuring the schedule of crewmembers to include
RSPs, and encouraging individuals to participate in them. While various
circumstances make the latter possible, it is likely that the skills and efforts of
expedition leaders are especially important. This would be the case whether
those leaders promote these rituals based on their intuitive understanding of
what is needed in this situation, past experiences, or more deliberate evalu-
ation and planning.

Special Collective Ritual Events

Several other concepts are also potentially useful for understanding what happens in expeditions. One such idea deals with special collective events, a type of social experience that can strongly affect individuals and groups. Such events occur in all kinds of societies, both past and present. Examples of these kinds of events include religious ceremonies, holiday celebrations, special family dinners, festivities, reunions, national commemorations, ethnic or community festivals, political rallies, military marches and celebrations, tributes to famous persons or important historical events, weddings, sporting events, pep rallies in high schools and colleges, musical concerts, public gatherings promoting a cause, and so on. As this long list of examples shows this type of event is quite common, can take place in groups of different sizes, and may take many forms.

A *special collective ritual event* is defined in terms of four key features (Knottnerus 2010; 2014a; 2014b). To begin with this type of social phenomenon is clearly demarcated and separated from normal, everyday activities. The nature of the separation can be quite extensive and clear-cut involving, for instance, markers and instructions that stipulate where and when the event should take place and its importance and meaning. It is considered to be a distinctive social behavior set aside from daily pursuits.

Second, such events occur in a standardized or regularized manner. They are typically engaged in on a periodic basis whether that involves a fixed time schedule, or their performance being linked to other occurrences such as a party celebrating a team's victory in a playoff, a military ceremony marking the completion of training, a pep rally preceding a ball game, or a celebration at NASA because of a successful mission.

Third, special collective ritual events involve to differing degrees stylized activities. In these events people engage in conduct that is quite noticeable due to its clearly recognized style or form. Examples of such activities include preaching a sermon, marching, making a toast, dancing, singing, praying, speech making, making vows or swearing oaths, presenting awards to individuals, recreational/sporting behaviors, and parading.

Finally, this kind of event involves multiple actors. What's more there is normally a belief and expectation among people that these occasions are collectively engaged in. I would stress two further points about this feature. The number of actors participating in such events can greatly differ ranging from tens of thousands of persons to just a handful of individuals. Moreover, a special case may exist in which a single person might engage in activities related to the collective event. It is possible that an individual who is isolated from others (e.g., a prisoner in solitary confinement in a POW camp or a penitentiary or a member of an expedition who is separated from other crewmembers) may celebrate the event and perform some version of the ritual, although in a modified manner. A simple example would be a person who reproduces part of a religious ritual such as reciting verses or singing songs, which had previously occurred within a group.

These four features outline the key characteristics of a special collective ritual event. Just as such phenomena are found throughout our ordinary lives they may also arise in expeditions. Actually, they may be engaged in by crewmembers in many different ways and quite often. Or they may not.

Whether they do or do not occur is an empirical question we shall examine. Furthermore, whether their presence or absence has important consequences for expeditions is another question we will address. In looking at this issue we will give particular attention to whether such events impact the shared emotions of crewmembers and if these emotional states influence the quality of life for these persons, their relations with each other, and their integration within the group.

This concern takes us to the next concept, which focuses on such matters. While evidence for assessing this formulation may be limited in a good number of cases it is still essential that its key arguments be clearly delineated.

Collective Emotions

The model of how special ritual collective events can influence people's emotions and commitment to a group focuses on four factors, two of which are composed of two subparts (Knottnerus 2010; 2014b). The four factors are attention, interactional pace, interdependence, and resources.

Shared focus of attention of actors in a collective event is concerned with the extent people's attention is directed to certain objects. Individuals' focus of attention can range from a lack of attention to an extremely high level of concentration or awareness. And people can focus on many different objects such as other persons, logos, or music. The key idea is that as the focus of attention of actors increases, the strength of the collective emotions shared by those in the event also intensifies. The nature of the emotions can also differ depending on what is focused on, e.g., feelings of satisfaction, joy, pride, delight, or disgust, anger, or animosity. Many characteristics of individuals and situational features can influence people's degree of attention ranging from the physical setting of the ritual event to the presentational style of a speaker, for instance a religious service or political rally.

Interactional pace of a collective ritual event deals with the degree to which people are involved in a series of interrelated acts and the nature of the repetitive or recurrent acts. It is a product of both: (a) rate of interaction and (b) if there is a rhythmic motion to their physical movements. Rate of interaction involves the speed, pace, or frequency with which people interact. The speed or pace that acts occur between individuals in collective events can greatly differ and may be measured in extremely small intervals such as seconds.

Rhythmic motion is concerned with whether and to what degree physical movements or actions recur in a consistent manner. Physical movements in a collective episode may exhibit a patterned and regularized quality in which the acts of individuals complement and coordinate with each other.

When the interactional pace increases in terms of rate of interaction and/or rhythmic motion, the strength of people's emotions increases. For instance, in a spirited party or celebratory event people may sing, dance, and toast each other at a very fast pace and in highly coordinated manner which increases the positive feelings group members have for each other.

Interdependence of actors deals with (a) the relative occurrence or distribution of acts by participants in a collective event and (b) how differentiated the actions are that are needed to carry out the collective episode. Stated somewhat differently interdependence of actors in regard to the first subcomponent concerns the degree to which actors may or may not be contributing to the ritual performance. For instance, only a single person or a handful of individuals may literally participate in the ritual act while all others only observe the performance. In contrast everyone who is present may actively participate in the shared action.

The second dimension of interdependence of actors is concerned with the level of complexity of different actions that comprise the collective episode. Here, the range of possibilities goes from only one or a few activities that people need to engage in to conduct the collective event to numerous activities that are needed to successfully produce the episode. In the latter case the collective occurrence involves a more complex endeavor where various specialized actions are needed to carry out the event.

In essence, the more people contribute to and are participants in the collective episode, the greater is the emotional impact of the event, and the higher the number of activities and complexity of actions, the greater is the sense or feelings of collective dependence and co-participation or co-involvement among individuals. What this results in is the heightened impact of the collective event on participants' emotions.

For example, in a simple collective assembly such as a political rally or a church service, a passive audience may merely observe the actions of one individual, i.e., a speaker, and only a few practices are engaged in during the collective episode such as clapping, reciting, or praying. In contrast, everyone attending the event may contribute to the collective episode by engaging in a number of activities such as singing, praying, chanting, reciting, dancing, swaying together, clapping, making pledges or donations, cheering, and/or making gestures. And in the latter case many different specialized practices such as these and others, e.g., carrying and waving banners and flags, bands playing music, large visual displays, filming the event, managing the lighting and sound systems, and/or parading may be critical for carrying out the collective event whether it be a mega-church service or mass rally. A famous example of the latter would be the Nuremberg rallies orchestrated by the Nazi party and Adolf Hitler. Here the collectively engaged in and multifaceted event had a powerful effect on participants' shared emotions.

The fourth factor, *resources*, refers to materials needed to perform rituals that are available to actors. Collective events require resources; they are indispensable to the enactment of such occurrences whether that involves human or non-human resources.

One of the most important human resources deals with the degree to which actors are *co-present* or visible to each other. The higher the likelihood that all individuals participating in the collective episode are seen by and cognizant of each other, the greater is their awareness that they are part of the shared exercise. What this results in is a greater effect of the shared activity on individuals, particularly their feelings about the collective experience. In other words, co-presence engenders the perception that there is support from and a consensus with one's companions, what could also be referred to as legitimation from group members. This sense of social endorsement or validation leads to increased feelings of satisfaction, confidence, and assurance about the collective episode. Of course, we should remember that co-presence and the feelings it creates occur in many kinds of groups, which are of quite different sizes. The fundamental dynamics are the same whether we are talking about thousands of people within an ancient Roman circular arena who are acutely aware of each other or a small number of crewmembers in an expedition who collectively engage in some activity in a small, confined space on a ship where an awareness of the presence of others is inescapable.

So too, many types of non-human resources can affect the production of a ritual event. The materials used to stage collective episodes could include clothing, costumes, written materials, food, musical instruments, signs, decorations, vehicles, or pictures. We should also recognize how important the physical setting can be. The structure of the physical environment can intensify the ritual performance and actors' awareness, as is the case in an arena, which focuses a crowd's attention on a center stage while also amplifying a sense of co-presence. And, as just noted, the same effect could also be obtained in a small physical space within for instance a sailing ship or research station where tight quarters intensify people's awareness of all participants in a collective episode. Finally, technology, machines, and different types of equipment can be used to enhance collective performances. Examples would range from the use of large televisions, digital imagery, and sound and lighting systems during the event to dog sleds that are used to construct a site for the ritual event or to actually carry out the activity.

These four factors, and their subparts – focus of attention, interactional pace (rate of interaction & rhythmic motion), interdependence of actors (differential/equal contribution of actors & degree of complexity of actions), and resources (human & non-human) – can vary. These elements influence the emotional intensity of people in a collective event.

Emotional intensity refers to the strength of emotions determined by these four elements. The nature of these emotional states can differ depending on the kind of ritual event, e.g., recreational, religious, or celebratory. These ritual dynamics influence the strength of positive or negative emotional states such as being aroused, delighted, excited, joyful, reverent, hopeful, contented, and satisfied or anger and hate.

Finally, the strengthening of emotional intensity among group members leads to higher levels of commitment to the practices engaged in by people during collective events and the beliefs expressed through these behaviors. Many kinds of emotions such as the ones just mentioned can enhance actors' commitment to the ritualized actions and beliefs of the group during the collective episode. This condition then impacts individuals' overall loyalty and dedication to the group. In other words, increased commitment to practices and beliefs during the event results in enhanced devotion to the group and strengthened social bonds.

In sum, the intensity of actors' collective emotions and consequent devotion to shared ritual practices, beliefs, and the group are influenced by four factors and their components. As these factors increase, the higher the emotional intensity, which strengthens commitment to ritualized practices and beliefs. This in turn increases dedication to and social solidarity among group members.

If special collective events occur during expeditions this could have major consequences for these missions. Such events could impact the shared emotions of crewmembers, which would then influence their interpersonal relations with each other, their morale, and their commitment to the group. This is another issue that warrants examination assuming the existence of relevant information.

Collective Pride

One final concept deserves our attention. Ritual events can have a powerful impact on the development of collective pride among group members. *Collective pride* refers to a high or inordinate opinion of a group's dignity, importance, merit, and/or superiority, grounded in group members' collectively shared emotions, cognitions, and practices (Knottnerus 2014a). In other words, these are shared beliefs people hold about the value of their group, beliefs that are to varying degrees emotionally charged and which play out through their actions with others.

Defined in this manner, collective pride can have various meanings that are determined by the characteristics of the groups that are emphasized. Value or importance could, for instance, concern the background of a group, the lifestyle of its members, their unique views and beliefs, or more encompassing qualities such as the moral purity or worth of the collective body. Collective pride can also involve all sorts of social units ranging from small-scale groups such as teams or clubs – more micro levels in society – to larger social bodies such as formal organizations or entire nations – more macro levels. Moreover, collective pride can be displayed in both positive and negative ways. It could involve realistic views about a group's merit or value based upon, for example, some accomplishment. In marked contrast, excessive, unwarranted claims of superiority might result in the vilification and oppression of others. Collective pride can, therefore, involve both productive, beneficial group practices and harmful, destructive forms of behavior.

Returning to the previous discussion of collective events and shared feelings these ideas focus on how collective episodes shape emotions, group commitment, and integration. What is not addressed, however, are the distinctive qualities of the beliefs related to the group and its ritual practices. The nature of these beliefs can have a powerful impact on the growth of collective pride.

Simply stated, the more beliefs emphasize the group itself, the more likely it is that collective pride will develop among group members. Compare, for instance, a celebration focused on a specific occasion such as a wedding, a baby shower, or a birthday to the celebration or commemoration of the group itself as in a tribute to a nation's past and its heroes, a rally or convention for a political party, a pep rally in a high school, or a memorial service in a military unit for its fallen comrades and the legacy of that unit. In the latter cases, the collective event and rituals explicitly stress the group and beliefs about the positive qualities of that social body. By directing attention to the social entity and its desirable attributes, the emotional intensity of the event amplifies people's beliefs about and sense of collective worth or pride in the group.

This idea is captured in the concept *reflexive ritual practice (RRP)*, which stands for a collective ritual that is directed towards the group performing the ritual (Knottnerus 2014a). *RRPs* focus on or are turned back upon the group that engages in the ritual activity and usually include beliefs about the qualities of the social entity such as its worth or goodness.

Building upon the model of special ritual collective events, increases in emotional intensity – resulting from focus or attention, interactional pace, interdependence of actors, and resources – leads to enhanced commitment to the rituals engaged in by individuals in collective events and the symbolic themes expressed by these activities. Emotional states such as happiness, excitement, or satisfaction energize and strengthen people's dedication to the practices and beliefs of the group during the collective episode. If these activities and the event as a whole include RRPs, actors' fervor and commitment to the rituals' beliefs about the qualities of the group are intensified. These heightened positive beliefs and emotions based on reflexive rituals amplify people's opinions about the value or merit of their group. Thus, collective pride is enhanced. Once this occurs, the reinforced collective pride of individuals may then influence the dynamics of the social unit resulting in, for example, an increase in cohesion among its members.

In regard to expeditions, it is possible that some crews come to share a sense of pride for themselves and the mission they are a part of. We will, therefore, look for evidence, which demonstrates this is the case, or at the very least, evidence that shows collective events and reflexive rituals exist which could be a source for collective pride among crewmembers.

Success or Failure of Expedition Crews

Of course, a key issue deals with the actual impact of reritualization on people. That is why a major focus of this study is to assess if rituals have an appreciable effect on the relative success or failure of expeditions defined in terms of the social psychological condition of crews.

For this reason, the social environment of expeditions is conceptualized in terms of two broad categories, which are comprised of a variety of features. These categories serve as indicators of overall social integration along with related attributes emphasized by different social psychological and sociological traditions ranging from research on group processes to scholars such as Emile Durkheim.

I would stress that when speaking of the success and failure of expeditions we are referring to the personal and interpersonal condition of the crews and not necessarily whether an expedition achieved its formal objectives. These are separate issues. For instance, there were expeditions that failed to reach their stated goal but were successful in terms of the quality of group ties and frame of mind of crewmembers. At the same time there were expeditions in which clear social psychological deficiencies existed which contributed to the failure of the missions.

Various attributes denote successful and failed conditions among crews. Generally, these features involve the two broad categories of strong versus weak or nonexistent social ties. More precisely strong and weak social ties signify the six contrasting qualities of solidarity/conflict, cooperation/non-cooperation, communication/lack of communication, positive emotions/negative emotions, high/low morale, and effective/ineffective leadership. These social psychological factors encompass various characteristics and descriptors, which can overlap to varying degrees with each other, e.g., solidarity/conflict, harmony/discord, cohesion/disintegration, or unity/disarray. Taken as a whole the six contrasting qualities convey a strong idea about the nature of the experiences and social dynamics of crewmembers.

These indicators will be elaborated on as we examine accounts of different expeditions. In doing so this should provide not just a clearer understanding about what they conceptually refer to but also a much richer description of these conditions and a greater appreciation for how they are realized in the day-to-day lives of individuals. The six pairs of social psychological indicators representing the two categories of successful and failed social conditions are summarized in Table 2.1.

TABLE 2.1 Social psychological indicators

Signs of successful conditions (strong social ties)	Signs of failed conditions (weak social ties)
Social solidarity, cohesion, integration, harmony	Social conflict, discord, disagreement, disunity
Cooperation	Non-cooperation
Communication	Lack of communication
Positive emotions	Negative emotions
High morale	Low morale
Effective leadership	Ineffective leadership

The Impact of Rituals

Lastly, rituals can have a profound impact on humans. There are a number of ways that RSPs are meaningful and of value. Previously I (Knottnerus [2011] 2016):128–130) identified nine ways rituals impact people. Here I reiterate these points but also expand the discussion and identify 11 ways rituals can affect individuals and groups. I am not, however, suggesting that every ritual activity is significant in each of these ways. RPSs vary enormously in their nature and form, the circumstances in which they occur, and how they are relevant to people. By highlighting these broad points now, we should be better able to appreciate the different ways rituals may have impacted the groups we will be examining.

The 11 ways, not listed in any order of importance, that rituals can impact people are as follows. First, RSPs facilitate social relations and communication. They provide a medium for interaction and socializing and in doing so foster social bonds and cohesion among individuals and group members. Second, RSPs provide guidelines that help structure or pattern social dynamics and personal behaviors. This produces a stability and order in human behavior and all kinds of groups. Third, rituals create a sense of familiarity and closeness to previous experiences and occasions of both a personal and social nature. By enacting previously engaged in RSPs individuals create a sense of grounding and security in their lives. Fourth, rituals give rise to and strengthen religious beliefs, commitment, trust, and feelings of fulfillment as has been long noted by a number of sociologists and other social scientists. These religious practices frequently involve collective experiences, which also reinforce social ties.

Fifth, rituals and ritual events can generate different emotions such as elation, joy, or reverence. Oftentimes rituals, which are intensely shared with others, create excitement, satisfaction, and happiness; they involve collective emotions of a positive nature. At the same time, some RSPs can engender shared emotional states that have much more negative connotations such as hate, contempt, disgust, or prejudice. Sixth, RSPs may create a sense of moral worth or purity. While this moral state may stem from a religious experience it does not have to; it can be attained through many different kinds of practices that are not necessarily religious in nature. There can also be both positive and negative dimensions to this because a sense of moral purity may be realized through a dualistic mode of thought in which an opposite value such as moral impurity or worthlessness is projected onto another person or group. Seventh, ritual practices can enable group members to achieve a feeling of collective pride. RSPs can play a key role in the development of an elevated opinion about the importance or superiority of a group. These shared beliefs are usually emotionally charged and are expressed through our actions with others.

Eighth, rituals may contribute to the development, maintenance, and transformation of identity. By engaging in certain kinds of rituals people can undergo major changes in how they see and define themselves. At the same time rituals may contribute to the development and preservation of many different types of social identity ranging from organizational identity or ethnic identity to the collective identity of smaller groups such as athletic teams or military units. Ninth, RSPs can create a sense of control and discipline. By enacting a ritual, sometimes in opposition to another group, competitor, or enemy, or in the face of very difficult obstacles, individuals acquire a sense of empowerment. A person knows that she or he can be in charge of and manage her or his behaviors, thoughts, and feelings.

Tenth, the performance of a ritual can create feelings of enthusiasm, dynamism, and vitality. Life affirming or energizing experiences involving rituals can be very motivating enhancing people's perception that they are living up to their full potential and are pushing themselves to the maximum. Eleventh, performing an RSP can be mentally or intellectually invigorating. The enactment of certain types of ritualized activities can have a powerful cognitive effect on people because they keep their minds active, attentive, alert, and require concentrated thought.

These are some of the ways rituals are consequential for people. At the same time RSPs very likely impact humans in other ways not identified here as will be elaborated upon in the final chapter. In examining the experiences of expedition crews, we should gain a greater understanding of how rituals affect these groups whether that involves some of the eleven just enumerated or possibly in other ways.

All of these ideas will guide our investigation of polar expeditions. Particular attention will be given to whether and how reritualization takes place and can help crewmembers cope with the stressful environment of an expedition. In examining these issues, we will give primary attention to the rank of RSPs and how these activities influence different groups.

We will further supplement this analysis by utilizing other concepts when appropriate to better understand different aspects of crewmembers' experiences. These ideas include the different types of RSPs engaged in, programs of ritualized practices, special collective ritual events, collective emotions, and collective pride.

In examining these issues, we will group expeditions into different types reflecting the absence or presence of RSPs and their effects. The organization of the next chapters will be based on these distinctions. Before we do that, however, we must first describe the research methods used to collect evidence about what life was like on these missions.

Methodology

In this section, several aspects of the research method will be discussed beginning with our analysis of the accounts for different cases.

Case and Data Selection

To conduct this study data comes from the accounts of expedition crewmembers as they were published particularly in memoirs and compilations along with other sources such as diaries and logbooks. Texts of this nature were used in order to gain insight into the day-to-day interactions of crews. Usually one text, i.e., account, was used to investigate the social dynamics of each expedition.

There are several rationales for using these types of sources in the current investigation (Prior 2004). Diaries and logbooks tend to be considered more historically authoritative in that they chronicle what occurred and when. Ship and expedition logbooks are usually official documents, so they are less likely to include personal notes and comments while the opposite is true regarding diaries. Memoirs are normally written after the fact and thus are subject to more careful considerations by the authors of what to include. As such they tend to be more cognizant of various social and historical circumstances that may affect people's lives (Bryant 2000). Memoirs also often draw upon sources such as diaries and logs written by the author and sometimes other crewmembers. Compilations are usually written to provide readers with a revealing and compelling account that is grounded in extensive research as is the case for those used in this study. The compilations examined here draw upon many different sources such as diaries, logs, congressional/governmental investigations, interviews, and historical studies. But they may also be subject to the temptation to selectively construct history.

While understanding these limitations, in a desire to include as much evidence as possible, we have drawn directly or indirectly upon these types of data. This approach is consistent with the perspective of Bryant (1994: 8) who argues that these kinds of documents "objectify and preserve significant collective experiences in the social rituals which regulate the rhythms of community life…" And in certain cases, it also allows us to make comparisons between different accounts of the same expedition, i.e., a triangulation approach (Lustick 1996). Overall, these kinds of accounts provide descriptions of actual occurrences on expeditions as well as introspective subjective accounts of the experiences of various individuals, which is extremely valuable for interpreting the more nuanced symbolic nature of crews' social dynamics (Lowenthal 1961; 1986).

Texts were selected via purposive sampling. This is not a random process. Rather attempts were made to be as systematic as possible in selecting texts. Using sampling guidelines for literary work provided by Glassner and Corzine (1982) and Van De Poel-Knottnerus and Knottnerus (1994) criteria were established regarding the body of literature to be examined. Several "famous" expeditions, some of which represented major milestones in Arctic and Antarctic exploration, along with a number of lesser-known missions were identified. Accounts from the mid-nineteenth to mid-twentieth centuries provided an array of viable choices for the investigation. This time span covers a significant period of time for examining polar expeditions in different conditions and settings while yet not being so different generalizations about the subject matter would be unreasonable. Nineteen expeditions were investigated. Leaders of the missions or other crewmembers wrote the memoirs examined here that were normally based on diaries and/or logs. Authors, not crewmembers, wrote the compilations and compiled large amounts of evidence from sources such as diaries, memoirs, interviews, and in certain cases official investigations. All the texts are published in English.

The large number of sources examined contains a rich and wide-ranging body of evidence about life on polar expeditions. We would also stress that the amount of time and effort needed to read, interpret, and analyze each text made it difficult to examine more than those chosen for this investigation. As it is the research took a number of years to complete.

Expeditions were conducted primarily for scientific, navigational, exploratory, and/or military purposes. Multiple cultures and nationalities such as indigenous peoples living in or near the polar region are represented in this sample, but the crews are mostly of European or American origin. Expeditions usually involved men although women were part of the crews in a few cases. Missions with culturally homogenous as well as culturally diverse crews were examined. Crews could include scientists, sailors, civilian adventurers, military personnel, physicians, native guides and hunters, and so on.

The time-period examined captures several eras of polar exploration, which can be generally identified as falling into three broad categories. The first category is Arctic expeditions prior to 1890 often searching for a possible "Northwest Passage" between the Atlantic and Pacific oceans or for lost expeditions. These expeditions also made inroads north, although, reaching the Pole was generally not the main rationale for the journeys. The second category involves sometimes quite famous expeditions racing to the North and South Poles, from approximately 1890–1917. The purpose of these expeditions was to either reach the poles or establish bases and waypoints to assist other missions in that quest. The third category is more modern expeditions from around 1918 until the mid-twentieth century. These missions were more scientifically oriented and were largely free of pressures to "be first" or reach some predetermined goal as a criterion for the success of the undertaking.

While most of the expeditions were studied using one account multiple sources were employed to examine several missions for quasi-triangulation purposes. This was done to increase confidence in and to further enrich the observations made about these missions (Lustick 1996).

Analytic Strategy

Analysis for this study was conducted as a literary ethnography, formally defined as "an intensive and systematic reading of an extensive range of texts possessing a clearly defined scope or subject matter" (Van De Poel-Knottnerus and Knottnerus 1994:70; 2002; Knottnerus and Van de Poel-Knottnerus 1999). Texts examined vary depending on the project and can range from memoirs, diaries, and compilations (of the sort examined in this study) to autobiographies, biographies, novels, or short stories. In accordance with this definition and strategy, selected texts were read with immersion and deep understanding as explicit goals.

A literary ethnography involves six steps. Briefly stated, one must first define the *scope of literary sources* to be studied. The researcher needs to determine what types of documents are available in terms of the topic that is of interest. In developing a clear idea about the range of documents to be examined one should have a clearly developed idea about the issue to be studied and the theoretical perspective(s) that is to be used.

Next, *reading and interpretation of works*, deals with how to read the literary sources. The researcher should engage in a focused concentration and reading of the texts identified in the first step. Probably many documents will need to be read. Generally, it is best not to focus only on well-known authors or accounts, but rather examine a wide array of literary sources as was done in this study.

Identification of textual themes concerns the discovery of various themes, which refer to individuals' experiences and the situation. The researcher now begins to recognize particular thematic elements that emerge in different sources. This is the first sign that certain themes appear in various works. Examples of thematic elements could include references to how people interact, personal experiences, and the ways people do or do not adjust to a difficult situation.

Once distinct themes have been identified the *classification of thematic elements* is established in which a classification scheme is developed that signifies or represents these themes. The investigator formally categorizes thematic elements that are referred to. The categorization scheme reflects the fact that one has uncovered distinct motifs in different texts. In other words, there are particular elements, which appear in various sources. One could think of this categorization scheme as a collection of empirical generalizations about, for instance, social relations, conditions, and ritual dynamics that are periodically if not repeatedly referred to in different documents. These generalizations provide a descriptive portrayal of individuals' experiences and circumstances.

The *development of analytical constructs* focuses on abstract concepts or generalizations about the qualities of the phenomena being investigated, i.e., the thematic generalizations categorized in the previous stage. Analytical constructs enhance our understanding of these phenomena because they are interpreted and explained in terms of abstract principles and formulations, i.e., theoretical approaches. These abstract constructs could draw upon theories and ideas in any of the social or behavioral sciences including, as in the present study, a social psychological/sociological perspective.

Lastly, a *contextual confirmation* should be carried out by rereading texts to determine whether the narratives correspond in a meaningful and accurate manner with the classification system and analytical constructs. The researcher may need to reexamine more than once some or all of the literary sources to be sure the identified motifs and abstract constructs are correct. By conducting a contextual confirmation confidence in the findings of the investigation is increased.

The outcome of the last step and the entire process is the creation of an ethnographic or composite portrait of the subject matter, such as the social world of a group whether that is a youth group in a school, a slave society on a plantation, or a crew on an expedition.

Based on the theoretical approach guiding this study we searched for especially three types of evidence. The first involves episodes of disruption and/or deritualization that caused people's ritualized behaviors to change. The second is any mention of RSPs and their effects, i.e., reritualization. The third are reflective statements. The latter involves commentaries about crews' experiences, rituals engaged in, and their possible impact on group members and their state of mind. The accounts examined here contain many references to disruption and deritualization and to differing degrees RSPs. When combined with reflective statements we gain a greater sense of the meaning and importance of people's experiences and the conditions they encountered.

All of the sources were read line-by-line, searching for references to these kinds of evidence along with any other pertinent information. The researchers recorded these findings.

Coding was, therefore, broadly guided by a three-part schema, which identified three key types of passages or text concerning: (1) disruption and deritualization, (2) reritualization and whether and how individuals coped with and adjusted to the situation, and (3) reflective statements about disruption, deritualization, RSPs, and the social conditions and morale of the crew.

Categories involving kinds of RSPs (e.g., recreational, scientific, hiking) and outcomes for the crews were developed and refined as texts were coded using an iterative approach (Strauss 1987). In regard to outcomes, assessments were made of social psychological consequences for the crews, i.e., morale and group dynamics. Several kinds of outcomes emerged involving low morale and limited cohesiveness, high morale and strong cohesiveness, and extremely high cohesion and morale.

Stated somewhat differently this research yielded a set of ethnographic "field notes" which informed our analysis. These "observations" (Glassner and Corzine 1982) were made in order to enable "sympathetic interpretation" (Wilson 1986: 137) of the symbolic nature of interactions amongst crewmembers. Thus, discussion of the data was driven by the investigators' qualitative descriptions of crews' experiences.

At the same time quantitative evidence is crucial to our analysis of ritual dynamics. This is especially the case when analyzing repetitiveness, which involves the frequency of ritual practices, one of the four factors that influence the rank of RSPs. Number of specific references in an account to different types of RPSs and other information about the degree ritual practices are engaged in are extremely valuable.

Guidelines for the Analysis of Documents

Several guidelines were used so that the researchers (two former doctoral students and I) would clearly understand what to look for when reading and analyzing documents. Five factors were formally identified:

1. Description of Expedition: Name of source, author, publisher, date of publication, dates, location, and goals of the mission, funding for the venture, nature of the environment and conditions faced by the crews, composition of the crew, how many people are involved, and other relevant information about the venture if available.
2. Disruption and Deritualization: Are descriptions provided of the disruptive experience involving isolation, dangers, confinement, high stress, etc., and their possible effects on actors? Are instances of deritualization referred to, i.e., are there descriptions of how the experience negatively affected people socially and/or personally?
3. Reritualization and RSPs: Do people engage in RSPs and if so, what is their rank and what kind of practices do they participate in, e.g., types of rituals or new or previously engaged in rituals? Who practices them, e.g., the entire crew, certain individuals, or subgroups? Do programs of RSPs exist and who promotes them? Is there any other relevant information about the rituals such as whether they generate collective emotions and/ or collective pride?
4. Reflective Statements: Are there commentary and appraisals about the mission concerning in particular the importance of RSPs for people and

how they impact crewmembers, how rituals may help people cope with conditions, or how by not engaging in RSPs people are less able to adjust to their situation?

5. Assessment of Success of Mission: Make an assessment about the social psychological success or lack of success of the mission, i.e., the signs of failed or successful social conditions. Do people cope with and adapt to the situation? Do they communicate and cooperate with each other, or are they unable to maintain normal, productive relations? Do the crew experience different kinds of difficulties or does social cohesion and positive emotions exist among crewmembers? What is the morale of the crew? How effective are the leaders?

Employing these guidelines, I read and analyzed all the expeditions. Dr. James Mason examined accounts of the expeditions in the beginning of the study while Dr. Kevin Johnson studied sources in the middle and later phases of the project. In addition to my investigating all the accounts Kevin and I reexamined, reviewed, and confirmed our findings.

Conclusion

In describing the experiences and social dynamics of different crews, the next three chapters are organized in terms of how poorly or well these groups functioned. We will begin by first looking at unsuccessful expeditions, followed by successful missions, and finally those ventures that are extremely successful.

Lastly, the first several expeditions will be described in greater depth in order to provide a fuller understanding of what polar conditions were actually like for the crews and the group dynamics on these missions. After that, in order to avoid an undue repetition of certain details, a good number of the remaining expeditions will be discussed in a more economical fashion.

3

FAILED EXPEDITIONS

*J. David Knottnerus, Kevin Johnson,
and James D. Mason*

All failed, successful, and extremely successful expeditions experienced disruptions and deritualization. The unique and threatening nature of these hazardous journeys created a sharp interruption in the daily rituals of crewmembers that shaped their normal pre-expedition lives. However, whether crews on these expeditions engaged in reritualization is another issue. In this chapter, we look at those expeditions where RSPs were generally the weakest and occurred to the smallest degree if at all.

Nine expeditions fall into the failed category. While they all involve polar explorations, they are quite varied in nature, size, dates of occurrence, backgrounds and nationality of crews, and whether or not they are well known. For clarity's sake we provide separate descriptions of each one. These discussions vary, however, partly because authors' accounts differ in their length and detail, and partly because the repetition of facts such as how expeditions are disruptive experiences is not always necessary and would become redundant. After our investigation of the first several ventures, which should provide a more in-depth understanding of what certain polar expeditions in the mid-nineteenth to mid-twentieth centuries period were like, we shorten our discussion of many of the remaining missions.

The Grinnell Voyage: A Troubled Search for the Franklin Expedition

Our examination of the first expedition is based on the journal written by Elisha Kent Kane, which is entitled *The United States Grinnell Expedition in Search of Sir John Franklin: A Personal Narrative* (1857). As the title implies the mission, captained by Lt. Edwin De Havan, and composed of an American

DOI: 10.4324/b23044-3

crew of 33 men, was one of many voyages, which went in search of the famous, missing Franklin expedition. Henry Grinnell, an American financier and philanthropist, essentially financed the venture. The expedition (1850–1851) was mostly located in the Canadian Polar Regions (and near Greenland). Kane, who kept detailed notes, served as the doctor for the crew.

This expedition like all the others examined here encountered many disruptions in the form of challenging problems faced by crewmembers. While every expedition is unique all the missions in this study experienced many of the same difficulties.

In the Grinnell expedition we find a wide variety of causes for these problems that negatively affected the crew's spirits. For example, long periods of time spent stuck in the polar ice in their ship, which could be for weeks or months, created boredom for all onboard. Kane states "I have continued my journal long enough to prove the wearing sameness of our days" during "our imprisonment. (p. 96)." Later, he says that the extracts from his journal "cannot escape the saddening monotony of the scenes that were about us (p. 210)." So too Kane observes that conditions such as "perpetual daylight," which occurred during parts of many trips, "disturbed me" and acted as "an unknown ... stimulus" adversely affecting sleep and eating (pp. 149, 150). And later in the journey he stresses how "we were all of us at this time undergoing changes unconsciously" involving their physical health and appearance, e.g., "shortness of breath" and a "peculiar waxy paleness" in their complexion (pp. 266, 267). Then Kane reports that: "Worse than this, our complete solitude, combined with permanent darkness, began to affect our *morale*. Men became moping, testy, and imaginative ... In a word, the health of our little company was broken in upon (p. 267)."

Weather also negatively impacted the crew. For instance, Kane emphasizes how the "rough sea," "gales(s)," and "fogs," prevented making observations that would help them determine the location of their ship (p. 152). Later he notes, "How very dismal every thing seems!" due to the driving snow and howling winds and how he is "In bed, reading or trying to read" while the ship is punished by an increasing "gale" (pp. 182, 183). Even minor routines are disrupted as evidenced in the author's statement that "in the little fireless cabin in which I now write, water and coffee are freezing (p. 184)."

Of course, the low temperatures of the polar regions created the aforementioned problem common to so many expeditions of their ships becoming trapped in ice, resulting in, amongst other things, constant monitoring and disrupted meals, work schedules, and even journal writing. In this regard Kane discusses how at one point in their sea journey they are trapped and drifting in a "moving wreck of ice-fields (pp. 248–251)." He goes on to describe how "the next four days were full of excitement and anxiety ... Every thing betokened a crisis" [as they prepared to abandon ship if the ice crushed their vessel] (pp. 248–251). And he stresses "The danger which surrounds us is so immediate that in the bustle of preparation for emergency I could not spend a moment upon my journal" … as they "endured a protracted succession of hazards (pp. 248–251)."

Moreover, we find on this and most other missions, crewmembers were subject to threats to their health involving various sorts of illnesses, injuries, and sometimes death. In the *Grinnell* expedition a major problem involved scurvy, a disease characterized by spongy gums, loosening of the teeth, and bleeding into the skin and mucous membranes along with other assorted physical problems caused by a lack of vitamin C. Actually, Kane states that: "It required strenuous and constant effort at washing, diet, and exercise to keep the scurvy at bay (p. 267)." Eight cases of scorbutic gums already required his attention along with one severe case of pneumonia. Then he (p. 271) states that the health of the crew is sinking "under the continued influences of darkness and cold" and "The symptoms of scurvy among the crew are still increasing and becoming more general (p. 267)." Faces are growing pale; strongmen pant for breath upon ascending a ladder, and an indolence akin to apathy seems to be creeping over us." Two days later he emphasizes how startled he is at the "traces, moral and physical, of our Arctic winter life … [one] can hardly realize the operation of the host of retarding influences that belong to a Polar night (pp. 272, 273)."

Additional observations are made about the health of the crew as seen in the following statements: "My scurvy patients …with all the care that it is possible to give them, are perhaps no worse; but pains in the joints, rheumatisms, coughs, loss of appetite, and general debility, extend over the whole company (p. 288)." Further on in the journey Kane warns that "An increased disposition to scurvy shows itself" affecting the men's gums and limbs along with their bodies becoming marked by blotches (p. 304). And these conditions involve not just the crew but also officers. Kane states "The scurvy has at last fairly extended to our own little body, the officers" who are experiencing "pains in the limbs and deepening soreness of the bones (p. 306)." The next day he points out how the scurvy continues to increase and that he is also sick, the effect of this condition being both mental and physical: "It is strangely depressing: a concentrated 'fresh cold' pain extends searchingly from top to toe. I am so stiff that it is only by an effort that I can walk the deck, and that limpingly (p. 308)." He also observes that a small number of men were "without ulcerated gum and blotched limbs (p. 311)." In other words, the large majority of the crew and all the officers including their physician became ill. And, despite periodic improvements and changes in their diets and exercise which were "enforced as part of our sanitary discipline … the disease [came back] with renewed and even exacerbated virulence (p. 325, 326)."

These are some of the problems faced by the expedition's crew. As we shall see, however, many polar missions experienced not only similar difficulties but also other kinds of troubles. In innumerable ways the lives of crewmembers were disrupted as they confronted the harrowing challenges that exploration of the Polar regions presented.

With that said, the key questions concern whether the crew engaged in ritual practices, the degree to which they did, the nature of these activities, and their impact on crewmembers. A careful reading of Kane's journal shows that RSPs were periodically engaged in, and they sometimes had a salutary effect. Several references to beneficial ritualized behaviors are referred to early in the expedition. Kane observes how they had "already begun to realize that power of adaptation to a new state of things ... We marked our day by its routine" as the crew fell into a pattern of always rising in the morning at the same time and eating at set times during the day (p. 35). On July 4, the crew also made "a festive effort" to celebrate their "American birth-day (p. 64)." And when stuck in ice early in their journey they one day organized sporting events such as "foot-races [and] football (p. 92)."

Later in the voyage, Kane reports that he has "been for some evenings giving lectures on topics of science, the atmosphere, the barometer ... to the crew" and that while "They are not a very intellectual audience ... they listen with apparent interest, and express themselves gratefully (pp. 239, 240)." Occasionally the crew also collectively held a special event that they enjoyed. Kane describes how the "crew determined to celebrate 'El regresado del sol ... a ... holy day ... [in which a crewman performed a] "theatrical exhibition ... followed by a pantomine (pp. 294–295)." A playbill "tackled on the main mast" described the production and Kane states that "The affair was altogether creditable ... every body enjoyed it" with the entertainment followed by the singing of a song and a round of rum (pp. 294–296). Furthermore, the doctor notes that both officers and crew tried to engage in a mid-day exercise on a daily basis unless storms and drifts prevented it (p. 310). While painful they played "football and sliding, followed by regular games of romps, leap-frog, and tumbling in the snow (p. 310)." Finally, on a couple of occasions Kane briefly mentions how on Sunday they held "religious service as usual" or a "divine service (pp. 245, 338)." Interestingly though the first mention of these services is not made until far into the expedition and no description is provided of them or how they may have affected the men.

Other than these ritualized activities, however, essentially the only other RSPs discussed involve activities engaged in by Kane, or occasionally he and one or two others. In fact, references to the latter far outweigh the aforementioned statements about ritual behaviors that might have had some beneficial effects or may have been enjoyable for the crew. In addition to his journal writing probably the most salient and repetitious activity engaged in by Kane was the repeated walks he took on land and ice when the opportunity presented itself. Repeated references are made to this ritualized behavior, which he speaks of in positive if not glowing terms. For instance, Kane on a hike of over five miles with two others describes how over "a long meadow of recent ice … with a fine bracing air, every nerve tingling with the exercise, and the hoary rime whitening your beard, you walk with a delightful sense of ease and enjoyment (p. 229)." Two pages later in his journal he states: "Tempted by the over-arching beauty of the sky, I started off this morning with Captain De Haven [Captain of the ship and expedition] on a walk of inspection shoreward (p. 231)." Another time he "Walked out to see the ice" and says he has "no change of words left to describe noonday" because both the sun and moon were in the sky. He then makes a very telling statement about how impressed he is with the view and how "I make my *four to six hours of daily walk* … [italics mine] (p. 238)."

Kane then goes on to make numerous references to this practice. For example he notes how on some of his walks he studies the topography of their "ice-island residence," he goes on a walk with another individual when the temperature is "–25" degrees, and he takes a several hours-long walk while describing how "I had a feeling of racy enjoyment as I found myself once more away from the ship, ranging among the floes, and watching the rivalry of day with night in the zenith (pp. 241, 261, 275)." Later, he mentions another walk with a crewman and on a different occasion how he becomes energized when he can "tramp away freely" on the floes and takes a walk with one of the ship's cooks who shares various stories with Kane (pp. 279, 308, 309). And on it goes as Kane refers to a "heavy walk," a walk with another crewman, a "comfortable" walk by himself, a lone walk "to be remembered," a walk at night "in search of open water," a "quiet day's walk," a "delightful tramp," a "long straight-forward march" with two others, and "pleasant … rambles on the ice (pp. 330, 339, 346, 347, 348, 396, 397, 418, 437, 438, 440)." Finally, confirming a key point already made he states, "I have been in the daily habit of taking long, solitary walks upon the ice, miles from the ship (p. 402)."

It is this personal, individual RSP, which stands out in his journal and his daily behavior. Clearly this practice played a central role in his life and in sustaining his spirits and sense of stability during the expedition. To a much lesser degree he also mentions another activity that he engaged in, which was reading (pp. 183, 236). But by far, walking and the near daily practice of diary writing, which often involved detailed descriptions of his and others' activities, were the two RSPs that provided a sense of meaning, direction, and focus for the doctor.

The remaining commentary to ritualized behaviors engaged in by the crew is most definitely of a different nature. Kane provides repeated accounts of activities which do not arouse and motivate crewmembers and that are essentially weak or of a mixed or ambivalent nature. References to these types of practices and the low morale of the men are what dominate his observations of those on board.

Besides commentary about the low morale of the crew, which was partly addressed when discussing disruption and deritualization, Kane makes many other statements of this sort. And these comments begin not long after the start of the mission. For instance, in mentioning how in August "Our prospects were far from cheery" due to the lack of progress made in the summer and eight months of bad weather to come, Kane indicates that the crew is fairly dejected (p. 105). Somewhat later as they begin to make more progress in their journey Kane does state "the tone of feeling rose among our little party (pp. 147, 148)." But not long after that as they encounter bad weather he declares: "How very dismal every thing seems! The snow is driven like sand ... obscuring the atmosphere with a white darkness. The wind, too, is howling ... (p. 182)." The unhappiness of the crew, or at least the officers, is also evident when the Captain decides to return home, thus, breaking a partnership with a British ship, which has been accompanying them. Kane points out "there was but one feeling among the officers of our little squadron, that of unmitigated regret that we were no longer to co-operate with our gallant associates under the sister flag," and then emphasizes how "saddened" they are (pp. 186, 187).

Lack of commentary in accounts can also be quite telling. In certain, other expeditions, as we shall see, major holidays become occasions for celebration. But when thanksgiving arrives on the Grinnell there is literally no mention by Kane of the day or any activities. And when religious services are held, a potentially highly ranked RSP, the couple references to these practices are totally lacking in spirit and positive content. For instance, Kane reports: "It is Sunday: we have had religious service as usual, and after it that relic of effete absurdity, the reading of the 'Rules and Regulations' (p. 245)." So too on another Sunday the only comment he makes is, this has been "a day of rest (p. 252)." No description is offered of what people did or could have done whether alone or collectively.

Concerning special occasions Kane's account of one Christmas is also revealing as seen in his referring to the "dreary intervals of leisure that heralded our Christmas (p. 257)." Later he provides more detail about what occurred on that day. On the one hand, the crew engaged in several different activities including a dinner, a play, singing, games, and gift-giving, which Kane describes as quite enjoyable. He describes a play which was so bad everyone felt it was very funny and says "... our Arctic theatre: it was one continual frolic from beginning to end (p. 269)." He also describes how they sang songs and had "a foot race in the morning over the midnight ice" and they "were very happy fellows (pp. 269, 270)." Yet, he begins his discussion of Christmas by saying "'Goodies we had galore; but that best of earthly blessings, the communion of loved sympathies, these Arctic cruisers had not (p. 268)." He then goes on to state: "It was curious to observe the depressing influences of each man's home thoughts, and absolutely saddening the effort of each man to impose upon his neighbor and be very boon and jolly. We joked incessantly, but badly, and laughed incessantly, but badly too ... (p. 268)."

In essence the doctor has very mixed feelings about the event, an ambivalence that is clearly expressed in his description. One is left with the strong impression that the feelings shared by all were at best muted and morale was not significantly improved, if at all.

The same also applies to New Years. Kane describes the event in no uncertain terms: "The first day of 1851 set in cold ... We celebrated it by an extra dinner, a plum-cake unfrosted for the occasion, and a couple of our residuary bottles of wine. But there was *no joy in our merriment: we were weary of the night* (italics mine), as those who watch for the morning (p. 275)." It is another weak collective ritual event that is lacking in spirit and cannot lift the morale of all those involved. A few days later he also refers to their engaging in "national festivities" and how the "evening brought the theatricals again, with extempore interludes" but emphasizes how it was "not thoroughly gladsome" and they "made the best of the occasion (p. 276)." Again, it was an event that was essentially lacking in happiness and cheer.

So, Kane too describes another event where the men struggle against harsh conditions and find little solace: "It is Washington's birthday, when 'hearts' should be glad; but we have no wine for the dinner-table and are too sick for artificial merriment without it. Our crew, however, good patriotic wretches, got up a theatrical performance ... (pp. 304, 305)." He then goes on to only describe the severity of the weather, −46 degrees outside the ship and −30 degrees inside, and how the condensation was so extreme they could hardly see the performers.

Even toward the end of the mission when they stop and visit a small Danish settlement, we find a similar state of mind among the group marked by a lack of emotion and enthusiasm. It was on July 4 and Kane says: "Our celebration was of the primitive order. We saluted the town ... and made an egg-nogg of eider eggs ... and, in a word, we all did our best to make the day differ from other days - which *attempt failed* [italics mine] (pp. 427, 428)." He then notes that two days later, on Sunday they attended a church service at the settlement but simply states: "The service consisted of a long-winded hymn, and a longer winded sermon (p. 428)." That commentary is then followed with the bland statement that they "put to sea in the afternoon (p. 428)."

Finally, additional observations point to the low morale of the men. Kane unequivocally states: "I have referred more than once already to the effects of the long-continued night on the health of our crowded ship's company. It was even more painful to notice its influence on their temper and spirits. Among the officers this was less observable (p. 290)." He then expresses the opinion that: "With the men, however, it was different. More deficient in the resources of education, and less restrained by conventional usages or the principle of honor from communicating to each other what they felt, all sympathized in the imaginary terrors which each one conjured up (p. 291)." Actually, this observation points to a division between the officers and the rest of the crew and how the former may not have exercised a great influence on the latter in regard to lifting their morale.

A little later in the journey the author himself admits how debilitating conditions are even for him. His journal entry reads: "Day very hazy and nothing to interrupt its monotony. It requires an effort to bear up against this solemn transit of unvarying time (p. 300)." He then proceeds to describe a typical day for the crew. Several comments deserve taking note of. After describing the cramped and extremely offensive smells of the area where all the men live and sleep, he then discusses how after their rising early in the morning and eating breakfast they go outside to work. Kane's only comment is: "One hour spent now in an attempt at study – vainly enough, poor devil! But he does try, and what little he does is done then (p. 302)." Then the officers and crew engage outside in their "anti-scorbutic exercise [exercises that are intended to fight the scurvy they suffer from]," a daily activity he refers to as "the usual foot-ball (p. 302)." These are exercises that are not so much done for their recreational benefit than for their presumably medical value.

After that comes dinner and the short afternoon quickly followed by night when Kane says they face the "saddening question, What has the day achieved? And then we stretch ourselves out under the hatches, and sleep to the music of our thirty odd room-mates (p. 302)." His only other comment concerns how they may hold their "arctic theatre" in the evening, a form of "cold merriment," which attempts with limited success to break the monotony (p. 303).

In conclusion, we find the Grinnell expedition experienced numerous difficulties which led to the loss of ritual practices that shaped the daily lives of crewmembers prior to this venture, i.e., deritualization. These problematic disruptions involved long periods of time marked by boredom and monotony, disturbing physical conditions such as constant daylight, and the weather ranging from severe storms to extremely low temperatures. The latter also resulted in the potential hazard of their ship being trapped and crushed by ice. Furthermore, various threats to the health of the crew existed which could lead to sickness and injury. In this particular mission, scurvy was the most conspicuous ailment that nearly all the crew suffered from. These conditions acted as a weight, which dampened people's spirits.

Some constructive rituals were periodically practiced by the crew, which on occasion had a beneficial impact on them. These included certain celebrations, some theatrical productions, exercises, and games on the ice. In terms of our analytic framework these helpful RSPs generally exhibited a moderate if not low degree of salience, some were repeated fairly often, while others less so, somewhat adequate resources enabled people to engage in these activities, and they were to a limited degree similar in meaning or form (i.e., homologousness) because they occasionally emphasized the theme of people joining together for some common purpose. Overall, they were of fairly low rank or importance.

Actually, much of the commentary unmistakably emphasizes the weak and diminished nature of ritual practices. And this includes the practices just mentioned because these activities were often described in a negative or ambivalent manner. Furthermore, the RSPs engaged in by crewmembers were fairly limited, usually involving ritualized celebrations, plays, and daily exercising. In other words, they were not very diverse. They also involved reconstituted rituals, i.e., activities that were, or could easily have been, practiced by participants prior to the expedition. Finally, while some evidence indicates solidarity, cooperation, and communication among crewmembers, e.g., theatrical productions especially among a more dedicated subgroup of crewmen, these conditions were limited in nature, scope, and the persons directly involved.

In fact, Kane's more individualized rituals are referred to most often and are described in much more positive terms. Essentially, they involved his numerous walks outdoors and his daily journal writing. For him and perhaps those few others who accompanied him on different occasions these RSPs were of a higher rank due to their prominence or salience in his daily life, their frequent occurrence, his having the means and ability to engage in these practices, and their focusing on his personal enjoyment (along with the walks providing observations and experiences for Kane to write about in his journal). They were extremely helpful because they energized and motivated him while occupying his time in a meaningful way. Of course, we should also appreciate that these personalized, oftentimes solitary practices, usually did not connect him to others, i.e., they partially isolated him from the crew.

All in all, evidence indicates the crew's morale ranged from low to moderate levels and the weakened, lower ranked rituals they engaged in, didn't significantly improve their mood or the quality of their interpersonal relations. This is especially so since there is no evidence to suggest that these practices and shared events such as celebratory gatherings, plays, or exercises generated strong, positive collective emotions, a sense of pride, or feelings of we-ness among crewmembers. This assessment is further supported by the fact that holidays and Sundays when religious services would have occurred usually weren't mentioned or if mentioned were not described, suggesting these rituals had little impact on individuals.

Lastly, in regard to possible programs of RSPs and collective events the most consequential one was the *personal program of RSPs* (Knottnerus [2011] 2016) Kane engaged in. The several rituals the crew participated in were much less significant and less helpful. Why was this so? It is hard to know for sure, but Kane provides one telling quote which may provide an answer: "Our captain is the best of sailors; but intent always on the primary objects and duties of his cruise, he is apt to forget or postpone a provident regard for those creature-comforts which have interest for others (p. 213)." While intent on the task at hand, apparently the Captain gave far less attention if any to the comfort, morale, or social conditions of the crew. In other words, he did not promote ritualized activities that could have had a constructive effect on all concerned. As a consequence, the leadership of the expedition was certainly not as effective as it could have been. This observation of Kane's is also somewhat ironic because in the next expedition we will examine where Kane is the Captain, he too doesn't appear to pay adequate attention to the quality of life and social dynamics of the crew.

In sum, this expedition was a marginal failure for the crew as a whole in terms of their ritual practices, morale, and group dynamics.

The Grinnell Expedition to Polar Greenland: Discord at Sea

Our investigation of the second Grinnell expedition is based upon an extensively researched compilation written by George W. Corner entitled *Doctor Kane of the Arctic Seas* (1972). This mission (1853–1855) occurred mainly in the Polar Regions around Northwest Greenland. Kane is now the Captain of the ship whose American crew was comprised of 18 men who were later joined in Greenland by an Eskimo youth and another man from Denmark. The purpose of the expedition, which again was sponsored by Henry Grinnell, was to search for the Franklin expedition and conduct geographical observations. This voyage was also problematic, although it differs in several important ways from the first Grinnell expedition.

While Kane and his crew accomplished a great deal in terms of exploring this part of the world and many crewmembers survived their arduous journey, they still experienced numerous problems and disruptions of their lives during their excursion. Some of these difficulties involved previously discussed problems such as bad weather, extremely low temperatures, extended periods of light or darkness depending on the time of year, and illnesses. For instance, Corner points out how at one point in the journey Kane suffered from a "mental deterioration that was affecting the whole party," which was "probably due to scurvy (p. 195)."

Even worse, however, were the brushes with death that occurred during the mission. In one instance, for example, a hiking party that went on a long journey encountered fogs, snow, and bitter cold which almost resulted in their freezing to death (pp. 150, 151). Fortunately, Kane led a rescue party that was able to save them. In other cases, the outcome was much worse. One man whom Kane knew longer than any other member of the crew died of tetanus (p. 155). Another individual perished because of erysipelas (a bacterial skin disease) after his foot was amputated (p. 155). And during a hazardous overland journey a crewmember sustained a severe internal injury while helping save a sledge, i.e., sled, which had fallen into the icy waters, and died a few days later. The others buried him in "a sort of trench and covered the corpse with rocks (p. 211)." Even the sledge dogs that were usually taken on expeditions for exploring land could suffer the same fate. Dozens of Newfoundlanders and Eskimo dogs perished during this mission due to the "darkness, cold, and half-starvation" and related physical illnesses (p. 146). More disruptive for the crew, however, was the state of their ship, the *Advance*, which became trapped in ice. Eventually the brig was abandoned, leaving only four men to take care of it, while the rest of the crew took to their sledges for a long dangerous journey by land (p. 208).

Still, the main questions deal with whether the men engaged in RSPs, the extent to which they did, what kind of practices they participated in, and their effect on everyone. A careful reading of Corner's account shows that some activities did occur which had a positive effect on certain individuals.

A few references are made to collective ritual events. In the beginning of the voyage Corner states that on July 4 "the crew enjoyed a special dinner, drinking toasts to Independence in porter and whiskey, and topping off with plum pudding" and they were given an extra rest period, although "serious trouble broke out" the following day when two crewmen were insubordinate to an officer because they were upset about an order (p. 130). Later in December, mention is also made of a special dinner for the crew of beef and grog (p. 144). And after various problems had surfaced among the men, to be discussed shortly, a rare collective event occurs in which all the officers and men including several troublesome individuals held a Christmas dinner and "forgot their discomforts" as they "opened their last bottle of champagne" and ate pork and beans which they imagined to be "turkey and roast beef (p. 194)." On only one other occasion is reference made to a holiday, Washington's Birthday, but there were "no spirits, animal or alcoholic, wherewith to share festivity" on that day (p. 197).

Corner also quotes from Kane's journal noting that they had "accustomed morning, and evening prayers" and Sunday prayers were regularly held for the crew (pp. 141, 145). Several references to praying are made although no commentary exists concerning how the crew felt about this (pp. 149, 174, 186, 210, 216). One other repeated behavior Kane engaged in with the crew is also mentioned. Being a great admirer of Tennyson the poet "Kane had read over and over again to himself and to his men Tennyson's two volumes of 1842 (p. 161)." Again, however, no information is available concerning the reactions of the crew to this practice.

Finally, reference is made only once to several RSPs the Captain encouraged among the men. The ship had become stuck in the ice for an entire winter and in November he attempted to keep the crew active by "sending them on reconnaissance trips and by promoting simple sports, such as ... football on the ice (p. 144)." Corner also reports "Indoors, card games, chess, and reading broke the monotony" and Kane and the surgeon read passages from the novel, *David Copperfield* to the crew (p. 144). He also tells us that the men held a "fancy-dress ball, and Kane started a newspaper (p. 144)." However, Corner immediately follows this commentary with the statements that "This was the *last forced show of homely festivity until the sun returned in the spring.* On Christmas Day there was *no holiday-making* (italics mine) (p. 144)." He then describes how the spirits of the men were extremely low in the months that followed (pp. 144–146). For nearly the whole winter these rituals simply ceased or had little impact on the crew, thus, failing to lift their morale.

In essence relatively few references to RSPs among the crew are found and when they are evidence strongly suggests these practices were quite weak. They were not particularly if at all salient, some occurred infrequently, and they did not exhibit a complementary (homologous) quality such as their celebrating the collective good and camaraderie of the group. Even those that occurred more frequently such as daily prayers or perhaps the poetry readings seemed to possess a low rank or level of importance. Simply stated little if any evidence indicates these limited, weak RSPs had a positive effect on the crew.

Actually, different statements emphasize the opposite, i.e., that these practices had no effect and possibly even a negative impact. For instance, after the ship had been stuck in the ice for 11 months Kane's prayers with the crew changed and expressed his "fading hope" about their venture (p. 174). Various events such as the previously mentioned celebration of Washington's Birthday, which was not a festive event, and even entries in Kane's journal attest to the weakened nature of these activities. On one occasion despite their usual Sunday prayers the men were "nervous and depressed," one crewman went without eating due to homesickness, and Kane himself felt a "heavy strain" as revealed in his diary (p. 186). And as already noted when they entered a long winter with their brig trapped in the ice, their Christmas was totally flat and lacking in celebration. From then on during the winter months there were no festivities. Not surprisingly we find no evidence of strong, positive collective emotions in their activities, any type of shared pride among the crew, or the development of any feelings of solidarity.

Indeed, it seems that as in the case of the first expedition the most effective rituals on board the ship involved Kane's own personal practices. These behaviors mainly involved the frequent, constantly referred to (by Corner) entries Kane made in his journal, oftentimes at night, and other related activities such as note taking and sketching different objects ranging from animals to a group of Eskimos (pp. 58, 169, 185, 215, 264). As in the first expedition, Kane's personal program of rituals appear to be the most consequential RSPs on this mission. While sometimes journal entries expressed his worries and fears about the expedition these almost daily practices were meaningful and fulfilling activities, which provided Kane a sense of focus and order in his life. However, due to the individualized nature of these practices they didn't involve the rest of the crew and, therefore, did not contribute to their morale, help improve relations with each other, or strengthen the stability of the group.

Finally, intertwined with comments revealing weak and relatively few rituals among crewmembers, aside from the Captain's pursuits, are ubiquitous statements about a host of problems such as the lack of social cohesion among all on board and the men's low spirits. From the very beginning of the journey to nearly the end of the mission references abound to the troubles experienced by the crew and officers (while returning home several comments suggest morale did improve but this period represents a very small portion of the voyage). Numerous examples of these problems can be cited. Indeed, so many instances of these difficulties exist, space does not allow to fully describe all of them.

For instance, at the very beginning of the journey two crewmen were "insubordinate" and used "disrespectful language" toward an officer because they were offended by an order he gave (p. 130). A few days later one of the men created a "disturbance in the seamen's mess" (p. 131). Not long after that Kane and another officer accused each other of "carelessness" in the handling of the ship (p. 135). Later a crewman "assaulted" another individual and the Captain punished him by restricting him below deck (p. 137). This is followed by reference to some of the officers becoming privately critical of Kane's decision-making (p. 138). Perhaps more ominously about six months into the journey the Captain placed a crewman in "solitary confinement" for "mutinous conduct, refusing to obey orders, threatening to kill our dogs, and other acts of insubordination" (p. 146).

From this point on the already troubled morale and solidarity of the crew precipitously declined. On a quite difficult two-man sledge trip (e.g., rough pack ice and snow blindness) an individual who had already caused problems on the ship became "exhausted and ... desperate" and wanted to quit (p. 164). Seemingly falling into a state of "madness" he tried to use a rifle to escape (p. 164). The other crewman, however, disarmed him and the man stopped his revolt. On the ship conditions were also bad as the crew faced the possibility that their vessel, which was stuck in the ice pack in August, would not be able to free itself for their second year in the Arctic. The crew was hungry and not well, fuel was low, and they lacked fresh provisions. As Corner states "Kane knew that other officers shared the same dread and that the men were beginning to murmur (pp. 169, 170)."

The situation culminated in a large number of the crew deciding to abandon the mission. While Kane was quite upset by this secession, they were provided with two boats and supplies (pp. 174–177). Later the men returned to the ship, although Kane remained quite upset by their actions and essentially did not forgive or forget what they did. And even while they were gone the Captain refers to the remaining crewmembers' lack of discipline (p. 183). He also observes how on Sunday, after prayers, the men were visibly "nervous and depressed" (p. 186).

Once the group who seceded had returned conditions did not improve. Misbehavior, insubordination, and a lack of discipline continued with dissension and tensions permeating the group. Kane felt he had to act and therefore punished the men who misbehaved. In one case for instance, he "armed himself

with a heavy belaying pin, wait[ed] for the man on deck, and after a brief inter-
view" bashed the individual's skull with it (p. 194). The man recovered and for
a while people behaved themselves and exhibited some degree of "harmony and
social comfort," but at the same time other references are made to a "general
mental deterioration" among the entire party, "increasing illness and mental
torpor," and another altercation between two crewmembers (pp. 195–197).

One last event provides dramatic evidence of the problems that arose on this
expedition. Following the previously discussed incidents Kane learned that certain persons had engaged in thievery and were plotting to take the dogs and a
sledge to go live with a group of Eskimos. Kane confronted the men including a
key instigator and "knocked him down and mauled him until he cried for mercy"
(p. 199). At a quickly convened court-martial the man declared he would follow
the Captain's orders, but within an hour had escaped on foot, i.e., deserted. A
few days later Kane and several men went on a deer hunt. They encountered the
deserter whom Kane fired his rifle at but missed. Somewhat later Kane was able
to capture the individual who was hiding at an Eskimo settlement.

In sum, interwoven with commentary about the infrequent and low ranked
RSPs of the crew (except for Kane himself) are ubiquitous statements about
people's low spirits, group divisions, disagreements, fights, plotting, thievery, insubordination, secession and desertion, mutiny, lack of discipline, Kane
beating men along with other severe forms of punishment, and so on. These
are the most dominant themes by far in the description of the experiences and
social dynamics of the crew.

Framed in terms of the analytic framework used in this investigation we
find a conspicuous presence of social conflict, i.e., lack of group cohesion,
an absence of cooperation much of the time among different crewmembers,
limited if not a total lack of communication among many individuals, and
negative emotions such as dissatisfaction, anger, mistrust, suspicion, and low
morale. We also find an absence of strong, positive emotions in their shared
activities and no sense of pride among group members.

Why was it the case that so few consequential rituals were enacted on this
mission, which contributed to the unfortunate conditions just referred to?
While the evidence does not enable us to answer this question with complete
certainty several factors stand out which very likely facilitated these developments. First of all, Kane suffered from health problems including seasickness
and rheumatic fever, which may have at times impaired his ability to perform
all of his duties (p. 125). Secondly, there is evidence of strains between some
of the officers and the Captain, as in the case of one man who Kane removed
from the position of executive officer because he felt he was too inexperienced
for the job (p. 126). The individual was upset about this demotion and believed
Kane had slighted him.

Furthermore, Kane "was shakily playing a new role - captain of his own
ship and commander of a raw, largely inexperienced crew (p. 127)." The

inexperience of so many people including the Captain most likely contributed to "problems of leadership and discipline" and undermined the enactment of shared RSPs (p. 127). Additionally poor decisions by Kane and others concerning the selection of the crew set the stage for troubles later in the journey. For instance, a number of crewmembers were "last-minute recruits" picked "up on the New York waterfront (p. 126)." Two of these individuals were involved in many of the previously discussed problems. Indeed, one of them was the deserter Kane eventually caught.

Finally, a serious question must be raised about the type and quality of rituals promoted by Kane during the voyage. The RSPs he supported were too few, often weak, and involved few collective events such as holiday celebrations and special social gatherings. The ritualized activities he promoted centered on prayers with the crew and reading poetry to the men. These Kane-centered rituals, which reflected what he most valued, were quite likely not as interesting, meaningful, or significant for the others. Actually, no evidence can be found which shows that the crew benefited from them.

When all is said and done the second Grinnell expedition was a striking example of a very troubled mission.

The Jeanette's Search for the North Pole: A Tragic Venture

The fate of our third polar expedition is reflected in the title of the account we draw from: *Hell on Ice: The Saga of the Jeanette* (1938) written by Edward Ellsberg. This source is unique because unlike all the others examined in this study it is a semi-fictionalized work, which describes the voyage of the ship, *Jeanette*. However, the format used by Ellsberg is grounded in an extensive and to our best knowledge rigorous account of the expedition based on available evidence, e.g., notes, logbooks, recollections of survivors, and published and unpublished journals. Furthermore, the sources he drew upon included an extensive set of records from a Naval inquiry and a Congressional Investigation. The author's goal was to present the most accurate possible description of the mission in a readable narrative form.

The objective of the Jeanette expedition (1879–1881) that was commanded by G.W. De Long was to search for the North Pole by sailing through the Behring Sea. The officers were mainly American as was the leader while the all-male crew was mixed including some Europeans, two Alaskan Indians, and one Chinese cook. As we shall see troubles arose among many of the 33 crewmembers shortly into the expedition. And ultimately most of the crew would die. Indeed, their journey provides a markedly vivid example of a failed expedition in several senses of the word, a venture that was both extremely hazardous and plagued by personal and social problems.

As Ellsberg states: "from the very beginning the officers at the Navy Yard [where preparations were made for the journey] regarded the *Jeanette* herself as unsuitable for a serious polar voyage, and this hardly led to complete harmony (p. 9)." These concerns were especially justified because once the expedition began it wasn't long before they encountered a major disruption of the mission. Aside from the usual problems associated with polar expeditions most serious was "an early season encounter with the icefields and at so low a latitude, [that] was a sad blow to our hopes of exploration (p. 67)."

So bad was the situation that early September "ended with the helpless *Jeanette* solidly frozen into the arctic ice pack" where she was to aimlessly drift (p. 75). The consequences were both social and physical in nature. As Ellsberg reports "That freezing into immobility of the *Jeanette* in so low a latitude, fell like an icy shower on the spirits of our wardroom mess, and from that day sociability vanished (p. 76)." What is more, the ship became physically frozen in the ice at a 9-degree tilt, which made everything problematic from walking on the icy deck to eating on a level wardroom table (pp. 76, 77, 80, 81).

Other kinds of disruptions also occurred, sometimes of a less threatening nature, such as the peaceful calm of a Sunday morning and inspection being disturbed by a potentially dangerous polar bear being spotted near the ship (pp. 88, 89). Other problems, however, were more serious. One officer went partially blind for unknown reasons while a little later into the journey two officers became sick, including the ship's medical doctor (pp. 158, 159, 184). Then one of the sailors reported to the Captain that a mutiny was afoot among the crew although it was quickly determined that was not the case because the sailor was "insane (pp. 203–206)." This episode was followed by a number of the crew becoming sick with a "general debility," cramps, and other symptoms because they were suffering from lead poisoning caused by solder in their tomato cans (pp. 238–241).

Worst of all, however, the ice surrounding their ship placed increasing pressure on the *Jeanette*. Finally, the timbers broke which caused a major leak. As the water flooded into the doomed vessel it sank into the sea (pp. 262–268). All the men moved onto the ice pack with some supplies and using several small boats struggled to get to open water so they could row to Siberia.

What followed was a long and arduous journey as they either dragged their boats across the ice or sailed from one ice pack to another. Increasingly the men showed "signs of physical breakdown" due to exhaustion, lack of food, and the wet cold (p. 323). Worse yet a gale struck while the three boats were on open water. One boat sank with all eight men lost while the other two became separated (pp. 335, 336). As their situation deteriorated one of the boats came upon a barren stretch of land, where many of the men lost their toes and the leader of this group, Melville, became temporarily unable to walk (pp. 361–452). Then after reaching a native village, Melville organized a search party with native guides and the healthiest survivors to find the other group, which included Captain De Long. Eventually they found him but he and all the others who had been on the third boat were dead (pp. 398–409).

The dire conditions they were subjected to resulted in physical and mental deterioration and ultimately their deaths. This is confirmed by entries in the captain's journal that was found by his body (pp. 410, 411). In it, he records how the crew was starving. One by one they died leaving only the captain who then passed.

With that said, the main question concerns whether the crew engaged in RSPs, the degree to which they did, what sort of practices they participated in, and their impact. A careful reading of Ellsberg's account shows that a limited number of RSPs did occur, which were often of a weak nature and in some cases literally ignored by many of the crew.

To begin with, several references are made to the "routine drone of the sea" involving recurring naval activities such as crewmembers conducting their watches, i.e., serving as lookouts (pp. 58, 59). Certainly, on this ship the "Habit and the law of the sea" was strong (p. 60). Reflecting this emphasis allusions are made several times to the captain or his officers conducting Sunday inspections (pp. 89, 91). Perhaps more important, however, the captain held religious services every Sunday. These services were required by naval regulations, but the crew was not required to attend. And they usually didn't. As Ellsberg notes, "except for the officers aboard, the captain's congregation was small (p. 92)." Later he points out "Two things only broke up our unvarying daily routine - Divine Service on Sunday, and the weekly issue ... on Wednesday of two ounces of rum per man (p. 216)." Getting the men to gather around the whiskey barrel was never a problem, but "not a single seaman" and only four officers attended the captain's service (p. 216). Actually, later in the journey only two men attended a Thanksgiving Sunday Service (p. 254). And even during the period when the crew is on pack ice or sailing on open seas in their smaller boats the captain would hold Sunday services, but they were still attended by only a handful of men.

On the ship after services would come the Sunday dinner, which receives very little attention in Ellsberg's account and is not described as having any special meaning or being particularly enjoyable for the men. Also, while the captain ordered a daily outdoor exercise routine to keep up the crew's health, only once is this activity referred to (p. 128). Yet the routine of ship life was extremely important to him. After the ship had sprung a serious leak and a number of the men developed various health problems, Ellsberg pointedly states: "Oddly enough through all this, after the first week's struggle with the leak, we continued our scientific and meteorological observations. The captain clung to that routine as to a lifeline, which perhaps to him mentally it was ... (p. 194)."

Shortly, after that Ellsberg turns to the journal of the captain where he describes the daily activities of the crew (pp. 207–209). The captain states: "All our books are read, our stories related; our games of chess, cards and checkers long discontinued (p. 208)." He describes the daily pattern of activity as basically rising in the morning, eating breakfast, smoking and talking some, attending to various mundane tasks, having dinner, more of the same along with attempts at napping, supper, kill time activities, and then to bed. He also mentions how their meals "have a sameness" which they "know by heart (p. 209)." And even though the routine of ship life is by all appearances quite important to him he prefaces his description of their daily lives with the following words:

> Discouraging, very ... There can be no greater wear and tear on a man's mind and patience than this life in the pack. The absolute monotony; the unchanging round of hours; the awakening to the same things and the same conditions that one saw just before losing one's self in sleep; the same faces; the same dogs; the same ice; the same conviction that tomorrow will be exactly the same as today, if not more disagreeable .... (p. 207)

In essence, the routine of ship life, which the captain tried to uphold, and which was quite important to him, e.g., being a very religious man he never waivered from holding weekly Sunday services, was not that appealing to or beneficial for the crew.

We should recognize that on a few occasions mention is made of attempts and non-attempts to hold special collective events and celebrations. Remarkably one date (December 22), which was a sign for half the winter night having passed and the shortest day of the year that was capped off with a beautiful spectacle of auroras, failed to generate any enthusiasm or celebration (p. 154). Days later Christmas Eve is bleakly described as "everything that traditionally Christmas Eve is not (p. 154)." And while the officers shared a bottle of fine whiskey they are described as a "glum group (p. 155)." Then, even though they had a very good dinner, the statement is made that "Christmas Day, mainly because it lasted longer, was even more dreary than Christmas Eve (p. 156)." The best the group could muster happened on New Years, when the captain gave the crew several bodies of brandy, which resulted in some comic entertainment among the men. Quickly after that, however, sober "misgivings" returned to everyone about the conditions they faced (p. 158).

The only other reference to special collective events occurs the *following* Christmas and New Years. The crew held a minstrel show on Christmas Eve, but "Of Christmas Day itself, the less said the better (p. 230)." They had run out of certain foods so, for instance, their mince pies were now made of pemmican. On New Year's Eve more minstrel shows were held, a little rum was drunk, and the captain tried to raise hopes with a "fine speech to the crew" about how they would be home by next "New Years (p. 230)." But "The only trouble with the speech was that no one, including the captain himself, in his heart really believed it" so they once again "soberly" pondered their future (pp. 230, 231). These are the only times any mention is made of collective events during the expedition.

Overall, we find ritualized practices such as religious services, routine work tasks, and collective events to be of low salience or prominence, not engaged in to any great degree, limited in shared meaning and form since they did not give emphasis, for instance, to the idea of crewmembers coming together for some common purpose or shared enjoyment, and at best possessing adequate resources which became more and more depleted as conditions worsened. These RSPs were of low rank, which meant they had little impact on the crew. Moreover, the few special events that did occur, which were usually of a subdued nature, failed to generate any strong emotions, pride, or shared sense of camaraderie.

Finally, interlaced with comments revealing weak and relatively few rituals are numerous reports about the problems of the crew involving conflict, disobedience, lack of cohesion, unhappiness, and low morale. References to such difficulties occur throughout the entire journey, several examples of which can be provided.

From the very beginning the ship's officers thought the ship was unsuitable for the polar voyage, which undermined social harmony. And when early on in the journey they encounter ice, this further dampens the morale of the men as seen in statements about how suppers and breakfasts became "somber" gatherings (pp. 67, 68). So too when one of the crew tries to create some humor through the use of puns he is quickly criticized by the others (pp. 69–71). At the same time discord between the captain and one particular crewmember also emerged after only ten days on the ice (pp. 97–100). An individual named Collins who was a meteorologist/scientist became extremely upset about his formal status as a seaman and not an officer. While others such as Melville tried to reason with him, he became increasingly angry believing the captain was persecuting him. What resulted were very emotional outbursts on Collin's part, arguments with the captain, and his withdrawing from many in the group.

Furthermore, after two month's "close association … [people] were beginning to get tired of each other's company (p. 103)." The normal conditions of isolation that accompanies life on a ship became greatly magnified on the *Jeanette* as they remained stuck in the ice in their "small cramped ship" seeing the same faces each day and experiencing the irritations that emerge between people living so closely together (p. 103). Equally serious, if not more so, was the fact that they had a "captain who socially had retired into himself" as he became tired of everyone's company (p. 103). As Ellsberg reports: "[it was] impossible to conceal, a mental despondency [that] … gripped our captain … (p. 103)." Indeed, the captain confides to Melville about the pressure he feels. Uncertain as to why the ship cannot free itself from the ice, worried about the conditions of the ship and crew, and agonizing about the lack of success so far of the expedition he tells his engineer that "it's hardly bearable (p. 119)!"

Meanwhile, conflict between the captain and Collins continues. His anger about not being formally classified as an officer grows and becomes further exacerbated when he feels personally mistreated by the captain because of his order that all hands undergo a monthly physical examination by the doctor (pp. 142–145). The discord spreads throughout the crew as evidenced by the comment concerning "increased ill-will in the mess [dining area], owing probably to the general state of ragged nerves (p. 218)." Collins angrily and mistakenly accuses the doctor of waking him in the morning by slamming a door and then becomes angry with Melville for some minor if not non-existent offence (pp. 219–222). The next day a similar incident occurs over Melville's supposed insults directed at Collins, which is then followed by the latter complaining to the captain about the engineer "plaguing him" by sing-ing Irish songs (p. 223). The captain's response is to order the engineer to stop singing these and any other songs, which seems unfair and of poor judgment. Even though it is his "only relaxation" and is very likely an RSP enjoyed by the crew as a whole he is forced to stop (p. 224).

From that point on the state of affairs between various persons further deteriorates. Melville never again, except to give an order, speaks to Collins. More generally increasingly strained relations are evident among the officers as reflected in the observation that "Our wardroom mess was now in a fine state for sociability (p. 225)." One man talked incessantly but no one listened to him, one wouldn't talk to anyone, several different pairs of individuals wouldn't talk to each other for various reasons, another was quite reticent, and the captain "weighted down by his responsibilities, felt compelled to maintain an extreme official reserve (p. 225)." Only the doctor and Melville were able to carry on a normal conversation. The statement that "subconsciously we felt and acted as if we *were* going to a funeral, only it was – ours!" provides an apt description of the social atmosphere (p. 225).

As for Collins, it all comes to a head after he fails to conduct some of his work duties and participate in the mandatory exercises for all crewmembers. Captain De Long confronts him and after he becomes highly belligerent, is arrested and told he will be court-martialed when they return to the United States (pp. 228, 229).

Finally, after their ship sinks and everyone is on the pack ice struggling to survive conflict and low morale continues to impair the group dynamics of the crew. Collins and another individual named Newcomb, who is described as a naturalist on the expedition fail to differing degrees to carry out their share of the work (pp. 292–294). Newcomb at one point starts complaining and arguing with a Lieutenant and when ordered to stop refuses. The Lieutenant immediately reports his conduct to the captain who places him under arrest and brings charges against him for a court-martial when they return home (pp. 295–302). Now two men are under arrest, both of whom continue to complain, argue, and disturb not only the captain but also other crewmem-bers, at a time when cooperation is most needed.

In conclusion, interlaced with observations about intermittent and weak RSPs are numerous references to individuals' poor state of mind, social divisions and disharmony, disagreements, insubordination, poor communication, lack of discipline, arrests, and future court-martials. These are the most dominant themes describing the group dynamics of the crew. We also find at best few displays of strong, positive emotions in collective activities when such events did occur which was infrequently and no references to any kind of collective pride. When RSPs are engaged in, they involve either routine work duties or other practices that generate little enthusiasm such as the religious services held by the Captain.

As for why few vibrant rituals were enacted on the expedition three factors stand out. First, the dire condition of the crew both before and after their ship sank undoubtedly created a great deal of stress, troubled emotions, physical and mental deterioration, and a loss of resources, which made it difficult to perform and enjoy different sorts of RSPs.

Second, dissatisfaction about the suitability of the ship for the voyage undermined from the very start social cohesion and morale. This was clearly apparent among the officers and certain others who expressed such concerns.

The third contributing factor involves the ineffective leadership of the captain and to some extent other officers. One of the main rituals on the boat involved the weekly Sunday services conducted by the captain, but few crewmembers actually participated in them throughout the entire journey. The other principal activity involved mundane work duties of the crew, but no evidence is provided which indicates they had an uplifting effect on individuals' spirits or relations with each other. And the captain did not facilitate to any great degree special events that could have improved people's emotional state.

Why did the captain promote RSPs that were often too weak or ignored and in the case of collective events too few and often muted in their effects? Several reasons come to mind. Quite likely and simply put, he probably didn't fully appreciate the value of these kinds of activities during a long and stressful voyage such as this one. He also became highly stressed and despondent which undermined his ability to be an effective leader as seen in various statements which reveal how troubled he was along with other evidence such as how he would sometimes shift back and forth between exercising authority in a highly formal or a very informal manner. Moreover, the large number of problems he encountered with various crewmembers undoubtedly contributed to how he felt and the challenge of encouraging better relations among the men.

Finally, he most likely overemphasized certain RSPs because of his own personal preferences. For example, being quite religious he concentrated on providing religious services to others even when most of the crew failed to respond and might have benefited from other kinds of rituals. The importance of such activities for him personally are poignantly seen in his last journal entries where as the members of his group died one by one, he would pray and hold a service for them. Engaging in these religious practices were in a sense a personal program of RSPs for the Captain. The problem, however, is that focusing on these practices to such an extent made it less likely that the crew would engage in other potentially meaningful RSPs.

There is no doubt the *Jeanette* was a failed expedition. Concerning the behaviors of crewmembers, ample evidence shows this to be the case. And assuredly the dire conditions the group faced contributed to this situation. But it's also important to appreciate that other expeditions have endured extremely difficult, deadly conditions while maintaining high morale and strong bonds among the crew. In the next two chapters we shall look at a number of voyages that faced similar conditions. In examining these expeditions, it will be instructive to compare how the *Jeannette* differed from some of these other undertakings such as the Shackleton expedition, which was extremely successful. In Chapters 4 and 5, we will examine such missions including a number of highly hazardous ventures. But for now, we turn to another troubled expedition.

The Karluk Mission in the Northern Seas: An Arctic Disaster

The Last Voyage of the Karluk: A Survivor's Memoir of Arctic Disaster was written by William Laird McKinlay (1999) who was hired in 1913 by the well-known explorer Vilhjalmur Stefansson to serve as a member of the crew on the *Karluk*, the lead ship of his Arctic expedition. McKinlay was a member of the scientific staff who acted as meteorologist and magnetician. The goal of the expedition, which was sponsored by the Canadian government, was to explore the Polar Region north of Alaska but very shortly after beginning its voyage the ship was destroyed by ice. Stefansson left the ship to continue the exploration on another vessel leaving the mixed crew, which included 22 men from America, Canada, and Scotland, one Eskimo woman, and her two Eskimo children, to manage on their own. For a year McKinlay and the rest of the crew struggled to survive in the perilous Arctic climate. Written years after the journey, McKinlay draws upon his daily log, other published reports including the captain's and other crewmembers' diaries, and memory to describe their harrowing and disastrous undertaking.

From the very beginning the mission was plagued by problems and people's concerns about the undertaking. During preparations for the journey McKinlay expresses "misgivings about the management [Stefansson's] of the expedition" and states there were "rumblings of discontent among the scientific staff, and real doubts were being expressed about the plan of the campaign – or lack of it ... (p. 11)." Even Captain Bartlett, the commander of the ship, was worried about how the boat was not mechanically up to sailing through loose ice (pp. 11, 12). Furthermore, Captain Bartlett who was hired at the last moment was not involved in the recruitment of crewmembers. This too was a concern because while they were able seaman and McKinlay admired certain individuals, he also had "grave doubts in many cases about the other qualities, which would be necessary for harmonious living in the kind of circumstances, which might face us in the north (p. 12)." For instance, one crewman was a drug addict, another had venereal disease, and several, against orders, had smuggled liquor on board the ship (p. 12).

The expedition was actually made up of three ships which also created a problem because "There seemed to be no one in charge to direct operations" involving the loading of supplies on the vessels (p. 14). This led some to "become extremely dissatisfied with the plans" and preparations and their not being aware of details regarding provisions for the journey (p. 15). What resulted Mckinlay reports was a "hectic scramble to finish the loading of the ships and … confusion" that was always met by the answer that things would be cleared up later in the trip. When the crew finally came on board the Karluk "there was utter confusion on deck, not a square inch of vacant space, disorderly piles of stores scattered around, giving the appearance of a neglected junk-yard (p. 17)." The upshot was that as they began their journey people, supplies, and equipment were on the wrong vessels (p. 19).

Within approximately a week's time the *Karluk* got temporarily stuck in ice. From that point on the vessel was either within the reach of ice or literally trapped in it. About a month and a half-later, Stefansson, the leader of the entire expeditionary effort, went hunting for caribou on the ice with a group of men. He, however, never returned to the *Karluk* and the ship broke loose from the ice and was carried away.

As the winter approached, various difficulties befell the expedition. By mid-October, preparations began for wintering the ship. Meals were cut down to two a day and a man stood watch all day and another man all night. Weather began to worsen as blizzards and gales periodically developed. One crewman also began to act "queerly (p. 64)." He wouldn't speak with anyone, wouldn't join in games, and refused whiskey and cigarettes. McKinlay hypothesizes this could have been due to the lack of sunlight, which is sometimes referred to as arctic sickness or seasonal affective disorder. But he also acknowledges that this may not have been the cause of his "strange behavior (p. 64)." Furthermore, the "economizing at meals" began. Concerning meals, the cook was actually a drug addict and was "even more scornful than his mates of 'them scientists'" on board the ship (p. 64).

In January, the inevitable occurred when the pressure of the surrounding ice on the ship became too great creating a ten-foot hole in the side of the vessel at which time the captain ordered everyone to abandon ship. Since, they knew the ship would eventually sink, a large amount of supplies had already been taken off the boat and several snow houses built for the crew. As the ship sank the last thing the Captain did was to play the funeral march by Chopin on his victrola before he stepped onto the ice (p. 67).

Almost immediately the group experienced a number of serious problems. A small party of four men headed out on the ice to reach a nearby island but didn't return. Other smaller parties subsequently went to search for the lost party in addition to going on shorter trips with dogs and sleds to gain experience in ice travel since almost the entire crew had no experience in surviving on the ice. These travels were fraught with dangers ranging from frostbite, freezing hands and feet, and falling into the icy waters to their supplies being ruined by the water. Eventually they realized that the party of four was lost, i.e., dead.

After that the rest of the group began an arduous trip over ice to reach land. Upon reaching Wrangel Island the captain decided he would try to go to Siberia where he could hopefully find help and organize a rescue mission. Realizing there were "strains and stresses that ran through the whole outfit" and a "bad feeling" between various persons he divided everyone into "four parties, each with its own allocation of supplies (p. 92)." The parties could settle where they wanted on the island. Then the captain departed. Any hope of their leaving the island depended on the long journey he was undertaking.

Conditions were extremely difficult and stressful for everyone on the island. In the winter months they lived in igloo-like structures and during warmer months in tents. The adverse conditions ranged from extreme cold and blizzards to a variety of illnesses and physical ailments. In the cold, windy weather they were oftentimes forced to remain inside nearly 24-hours a day "in cramped quarters, cold and wet and weak, [which] was the ultimate in misery (p. 96)" Moreover, their food supplies would only last a few months even though they could be stranded here for up to a year waiting for help. And hunting for food such as seal meat was a very uncertain enterprise. Food, it should be remembered was needed not just for humans but also their sledge dogs. Oil supplies were used to provide heat to keep warm and to cook the food, but they were also limited. Once the oil ran out after several months, they then had to rely on finding driftwood to use for fuel.

Eventually, many of the men began to die. In some cases, people's body not only gave out, but also their will to live (p. 114). And sometimes the cause of death was unclear. One particular illness, which afflicted many persons, involved extreme swelling of different parts of the body (p. 120). The nature of this illness was a mystery although McKinlay became convinced that the cause was the pemmican they were eating because it had not been prepared with the proper proportion of protein and fat. Too little fat and the result is protein poisoning. If that were the case this would be another example of how poor planning and mismanagement of the expedition led to tragic consequences. Indeed, McKinley believed that Stefansson was ultimately responsible for this state of affairs. Illness was not always, however, the primary cause of death. In one case a crewman was found dead in his tent due to a shot to his head with a revolver lying next to him (p. 136). While the exact reason can never be known, other crewmen considered this to be a possible case of suicide due to despair about their situation or some other unknown reason.

In the end, approximately half of those who started the expedition perished. Undoubtedly nearly all of the others if not everyone would have died if help had not reached them. Fortunately, Captain Bartlett made it to Siberia and organized a rescue mission to save the remaining crewmembers.

As before, the key question concerns whether the crew engaged in ritualized activities, the types of RSPs they engaged in, to what extent, and their effect on the group. A thorough reading of McKinlay's description of the expedition shows that some ritual practices of mixed value did occur on the ship but after it sank far fewer are reported except for the author himself.

On the initial leg of the voyage periodic mention is made of RSPs on the ship, although the rank of these practices, e.g., degree of repetitiveness and salience, is usually not high. For instance, it is simply stated that when they crossed the Arctic Circle, they "celebrated at dinner with a bottle of wine (p. 20)." References are also made to work rituals, particularly tasks carried out by the scientists such as one individual collecting samples of sea life from the ocean, crewmembers competing and gambling on who could take the quickest depth readings which were done by winding a wire, and sometimes helping the one scientist who was dredging the ocean looking for sea life (pp. 27, 38, 39). However, a close reading of this account also reveals that most of the scientists were oftentimes unable to carry out their work and these work rituals stood out because much of the time "monotony and frustration" shaped the lives of almost everyone (p. 27). In that regard the author states that only he and two other members of the scientific staff were able to maintain a "programme of work (p. 40)."

Several references are made to men participating in games such as chess and bridge in the evenings or sewing and reading in their bunks during the day because they were attempting to "ward off boredom (p. 36, see also p. 65)." On one or two occasions mention is also made of how everyone tried to find something to do, and their spirits were high, e.g., the author makes a medicine chest for the doctor. Otherwise, discussion of activities such as chess and a championship competition occur against the backdrop of how they "were at pains to devise every means of keeping boredom at bay (p. 50)."

A different ritualized behavior much more solitary in nature is also described by the author when he discusses how he would take walks on the ice any chance he had, and when the weather would not allow him to do so he would run "round and round the ship" or "walk and run up and down" the vessel until he was exhausted (p. 49). He explains how he would often go for walks during the day or night and enjoyed making observations of the Polar Regions (pp. 57, 61). He also states that occasionally he and the captain would have conversations about roses, gardening, and the problems of the working and upper classes (pp. 56, 57). Aside from the few times when he is accompanied by someone on a walk and the "interludes" with the Captain, no mention is made of others participating in these types of repeated activities (p. 57).

In only two instances do the crew as a whole engage in large collective events. One occasion involved preparations for and the celebrating of Christmas (pp. 57–59). Different sports were participated in including races and shot-putting in addition to the captain providing whiskey for the crew to drink. On Christmas day, the walls were decorated with international flags and Christmas greetings were painted on a large section of canvas. People were "laughing and joking" as they took part in these preparations. An elaborate Christmas feast was then held accompanied by more whiskey for toasts, which even included the teetotalers. This was followed by a selection of sweets and cigars and then smoking and listening to music on the gramophone.

While certainly a happy experience, the exceptional quality of this collective ritual is also quite evident when the author states this was "the first time in ages everyone was really looking forward" to such an event. Later he finishes his description of Christmas Day with the statement that by the end of the day "we felt we ought to be doing something festive, but the euphoria had gone … and … we crept off to bed (p. 59)."

The other event involved a football match that was held on New Year's Day (pp. 62, 63). It was preceded by singing, a bit of dancing, the sharing of whiskey, and a toast to the New Year. The game itself was a spirited, enjoyable affair with many, not too serious injuries. At the same time, McKinlay pointedly prefaces his description of these activities with a comment about how everyone was initially "affected by a more than usual lethargy (p. 62)." Both events clearly stood out from the normal state of affairs marked by boredom, listlessness, and low morale. Actually, these are the only two references to vibrant collective events during the entire expedition.

After the ship sinks and the men are living in four groups on Wrangel Island there are almost no references to RSPs that are social in nature. Only one reference is made to a "camp routine" where the trading of items such as fish and tea between individuals and camps continued to the very end when they are rescued (p. 150).

Other than that, the only commentary about rituals refers to those engaged in by the author and sometimes practices he shares with a few others. For instance, McKinlay says "I kept up my walking and running exercise" and "sewed and sewed" my old clothes (pp. 77, 78). Elsewhere, he refers to the "usual routine" of periodically constructing a snow/tent shelter and daily sleeping rituals of first taking care of their boots and wet and dry socks and changing their wet clothes before going to sleep (pp. 82, 83). Later, after he had gone to live with the Eskimos on the expedition, he refers to the "routine at mealtimes" which never varied; among other things this involved sitting around the dish of food and dipping each bite in a separate dish of seal oil (p. 147)." And another time the author discusses how "I kept busy on wet or foggy days trying to mend my clothing ... I was much worse off than the fellows in the other tent, *who had never been as active as me* [italics mine], so their clothing had never had to stand such hard usage (p. 142)."

Moreover, when he is recuperating from an illness, he describes a "daily routine" of walking for exercise, collecting driftwood, building a large fire, and drinking tea which while an effort and exhausting helped him to retain his *high spirits* [italics mine] (pp. 105, 106). He also mentions that he and another crewman "went for a walk, or rather, a crawl of three or four miles (p. 106)." And then McKinlay says, "I kept up my efforts to ward off the effect of the numbing, stupefying monotony" by chopping wood, cutting ice, strolling around the spit, and cooking "when there was anything to cook (p. 107)." He also reports that he went on another four-mile walk and during his "excursions" searching for wood became "keenly interested in the variety" of types of woods in this environment (p. 108).

Quite importantly on several occasions he also emphasizes the importance of reciting to himself Psalm 121 from the Bible which gave him "a feeling of elation, a lightheartedness (p. 106)." Later, when he remembers that psalm and the inscription "Thy will be done," his spirits are "lifted" and he becomes "much more contented (p. 145)."

All of these activities, which were frequently engaged in and quite salient, i.e., of high rank, constituted a personal program of RSPs for McKinlay. They are, however, nearly the only vibrant rituals referred to during the crew's extended stay on the island.

Finally, interspersed with references to relatively weak rituals among crewmembers, aside from the practices of McKinlay, are frequent comments about the depressed state of mind and social disharmony of the crew. They occur from the very beginning of the mission until its end. As already pointed out poor planning and decision-making, inadequate communication, and disorganization before and at the beginning of the expedition fostered confusion, misgivings, and discontent among the group and Captain. These conditions undoubtedly contributed to the poor spirits and social health of the crew early in their journey. In that regard McKinlay points out how "the seeds of dissension were being sown all the time (p. 24)." For instance, the men became "very annoyed" when Stefansson issued arctic clothes to the Eskimos who already had their own warm clothing while no arctic clothes were given to the rest of the crew (p. 24). This led to "some wild talk among the crew" and speaking "quite openly of deserting (p. 24)."

Furthermore, as already noted, life on the ship was marked by boredom due to the fact that scientists and other crewmen alike could not do their normal work due to their being trapped in the ice. This situation leads the author to offer the sobering assessment that the inability to accomplish anything created a "mood of depression in some of the men [which] worked against the kind of *camaraderie* that might have made life more bearable in such difficult circumstances (p. 41)." He also states that "As the days passed … we turned more and more in on ourselves, ever more dependent on the warmth and comfort inside the ship, and increasingly aware of the imperfections and deficiencies of the *Karluk* (p. 47)." For instance, he points out that the "saloon table … was too small to let us sit down at meals together" and there were not enough dishes, chairs, and lamps (pp. 47, 48). Such a problem is especially significant when we consider this from a ritual perspective. What such physical conditions did was make it difficult for everyone to gather together on a daily basis and engage in upbeat, shared emotional experiences that could have created feelings of companionship and solidarity.

As the ship continued to drift the crew also started to compare their voyage to the plight of the *Jeanette* when it was crushed by ice. This led to "an atmosphere of gloom and despondency (pp. 52–55)." In contrast McKinlay's morale remained high in part because he remained active studying the records of the *Jeanette* expedition in order to learn how to deal with their situation.

Ironically after the Karluk sinks the crew's morale is temporarily at its highest. The need to deal with many tasks such as conducting an inventory and arranging their supplies and equipment meant that they were now constantly busy instead of being bored with nothing to do. They were in "high spirits (p. 69)." Then the sun reappeared which led to a "feast, followed by a riotous evening of singing and dancing (p. 71)." Still, while there would be "other musical evenings" none matched the intensity of that event (p. 71). Essentially from that point on, relations between the crewmembers degenerated and their spirits further declined.

And existing problems between different crewmembers did not get any better once the captain decided to leave for Siberia in hopes of finding help. In light of "the strains and stresses that ran through the whole outfit," the "bad feeling" between two men in particular, and the troubled "relations" between various persons, the captain divided the crew into different parties to live on the island (p. 92). This was still not enough to prevent further troubles. McKinlay states that soon after the captain left "Already the team spirit … was deteriorating, and quarrels were breaking out over the sharing of food (p. 97)." For example, an argument "raged" with "much strong and obscene language" over very small differences in the amount of biscuits distributed among the four parties (pp. 97, 98). And "angry criticism" and "ill-feeling[s]" developed over the amount of oil used by different groups and how some used their tealeaves for smoking instead of drinking (p. 98). Actually, in reflecting upon individuals such as the cook "with his unbridled tongue and capacity for lying" the author observes, "At times it seemed to me that in our very mixed community we had all the seeds of future disaster (p. 99)." He then says that without the captain, "the misery and desperation of our situation multiplied every weakness, every quirk of personality, every flaw in character, a thousandfold (p. 99)."

Other references to conflict, anger, troubled relations, theft, lying, prejudice, and poor morale also confirm this observation. They include a story that a crewman and a male Eskimo wanted to keep "all the hunting spoils" which led to "threats of violence, even of shooting" by other crewmembers, the crewman disparagingly referring to Eskimos as "bloody Indians," the male Eskimo expressing fear of the crewman, and a "noisy quarrel" between two other men (pp. 108–110). Later a big argument arises over how many rounds of ammunition crewmembers could have, questions being raised about thievery and lying by different individuals, an angry meeting "with a flow of charges and counter-charges" and "loud and obscene" language about who would take care of the sick and forage for food, and one person (not the cook) calling the author a "bloody scientist (pp. 129, 130)." At another time it is revealed that several men, one being the crewman who presumably committed suicide, had been stealing food (p. 137). Other references are also made to "trouble over the sharing-out of meat (p. 141)." And the previously referred to crewman and male Eskimo continued to distrust each other (pp. 143, 144).

It is incidences such as these, which led the author at the end of his memoir to offer a very candid and telling assessment of the expedition. After their rescue and arrival in Nome, Alaska everyone was greeted as heroes, but he in contrast states "I could not see what had happened to us an anything but abject failure … (pp. 159, 160)." McKinlay, who later would serve as an officer in World War I, then poignantly sums up his feelings in the following manner:

> Not all the horrors of the Eastern Front … could blot out the memories of that year in the Arctic. The loyalty, the comradeship, the esprit de corps of my fellow officers and of the men it was my privilege to command, enabled us to survive the horrors of the war, and I realized that this was what had been entirely missing up north; it was the lack of real comradeship that had left the scars, not the physical rigours and hazards of the ice pack, nor the deprivations on Wrangel Island. (p. 161)

In sum, the crew on the *Karluk* engaged in some rituals such as cards, sewing, and reading. These RSPs were of a relatively low degree of salience and appeared to occur fairly often, i.e., they were of low to moderate rank. Only a few events involving strong collective emotions are referred to. And morale and social bonds were at best of a moderate nature. Actually, morale was mixed and was sometimes quite low due to boredom and frustration. After the ship sank and the captain left to get help almost no rituals were engaged in and cohesion and the spirits of the crew worsened. The exception was the author who had a regularly engaged in personal program of RSPs such as taking walks, work routines, and reciting psalms. In his case morale was decidedly better.

Several things contributed to the failed state of the mission. To a certain degree the causes are similar to what we found in the *Jeanette* expedition with three broad factors standing out. The dangerous conditions faced by the crew surely contributed to the troubles they experienced. Also, from the very start the mission was plagued by poor preparation, disorganization, and inadequate communication along with poor personnel selection, all of which fostered divisiveness, dissatisfaction, and frustration among the crew including scientists and officers. Lastly, less than effective leadership, and probably more importantly, missing leadership due to the Captain's quite understandably leaving to find help resulted in few, if any successful attempts to promote RSPs among the group while on the island.

As a concluding observation, we would emphasize that on some expeditions such as the *Karluk* and the *Jeanette* what we find is a compounding, escalating series of events and factors, which lead to the failed state of the missions, i.e., the array of causes and developments just outlined. In these cases, it becomes extremely difficult, if not impossible, to separate and temporally identify all the factors in a clear-cut manner. They can all occur simultaneously. But the key point here is that, however it plays out, the lack of strong, vital RSPs is a crucial part of the social dynamic, which significantly contributes to the expedition's failure in terms of morale and other social conditions.

We now turn to a different type of failed mission. With the groundwork established through our lengthy examinations of the first four expeditions our discussions of many of the remaining ventures will be more concise.

A Russian Expedition to Siberia: Life and Death in the Arctic

In the Land of White Death: An Epic Story of Survival in the Siberian Arctic was written by Valerian Albanov (2000) who served as the navigator onboard the *Santa Anna*, a Russian expedition (1912–1914), although details about who exactly financed it are difficult to determine. The goals of this mission were to search for hunting grounds, exploration, and possibly map a route from the Atlantic to the Pacific via a north-east passage. This remarkable, but often ignored mission in the Arctic ice north of Russia, was captained by Georgiy Brusilov and included one woman, a rare occurrence in the expeditions of this time. Unfortunately, the captain was overly confident if not totally naive about the dangers confronting the voyage while the crew was woefully unqualified for this sort of undertaking, with only five of the twenty-three crewmembers being "genuine sailors (p. *xi*)."

Shortly into the journey, the ship became trapped in ice. After being stuck in the ice for one and a half years, part of the crew decided to try walking south to safety. Albanov's account describes what happened when the ten men, who were led by Albanov, attempted to bring rescue to those who remained on the ship though many understood, but did not discuss, how staying on the vessel would likely result in death. What followed was an arduous 90-day journey over islands, frozen sea, and glaciers in which nearly all of the men died. Ultimately, only Albanov and one other crewman survived from the entire *Santa Anna* crew.

Not surprisingly numerous disruptions similar to those encountered in other expeditions confronted this mission such as the ship being trapped in ice for over a year. On the boat prior to Albanov's group departing all the crew suffered from scurvy in addition to fuel, coal, and wood beginning to run out which led to talk of mutiny (p. *xii*). Albanov's group experienced a wide variety of deadly problems including lack of fuel, limited food supplies, frigid temperatures, blizzards, food eaten cold and/or raw due to the shortage of fuel, and their sleep once being interrupted by a storm which blew the men into the water resulting in the drowning of two crewmen (pp. 139, 140). The deaths of the crew occurred in different ways and for different reasons, although the exact cause was not always clear. For instance, one man set out alone to explore a new route to take and never returned (pp. 52, 53). And later, two men fell victim to some type of fatal illness that created paralysis and eventually the loss of their mental capacities (pp. 124–126, 135, 136).

As to whether the men engaged in ritual practices during their journey, only a handful of RSPs are referred to having to do with eating and celebrating, i.e., enjoying some soup and a biscuit after traveling 60 miles from the ship, killing a bear and rejoicing with a dinner which temporarily raised their spirits, and celebrating Albanov's Saint's Day with a bear steak breakfast (pp. 54, 68, 76). Mention is also made of the burial for a highly respected sailor, although Albanov stresses that there was no service nor any kind of emotional response by anyone, i.e., a very dead ritual (p. 134).

The only other RSPs referred to that are repeated, salient, and important are those practiced by Albanov. In the evenings he regularly wrote in his diary and made "calculations of the day's navigational observations (p. 40)." And after spending hours in the kayak with his legs bent, he would always, no matter how exhausted he was, do "gymnastic exercises" to restore feeling in his legs (p. 136).

Intermixed with the limited number of references to rituals, except for those practiced by the author, are repeated comments from the beginning of the journey until its end about the morale and social condition of the group. A few remarks are made about the men's spirits being high, usually for a short-period of time, after some unexpected positive development. They involve the group being encouraged by a good dream of Albanov's, being able to shoot some seals for food, and people being in good spirits because they reached an island where food (eider birds and their eggs) and driftwood for making a fire are abundant (pp. 46, 47, 113). Reference is also made to how happy they are to discover an abandoned camp containing biscuits, canned meat, and smoking tobacco but this occurs at the end of the mission when only Albanov and another man are alive and their prospects for rescue are now greater (pp. 148, 153, 165).

All of the remaining commentary emphasizes the low spirits, lethargy, deviance, and unrest among group members along with Albanov's damning criticisms of them. These comments begin as soon as they begin their journey when the author describes how a three-day blizzard created difficult sleeping conditions in their cramped tent that led to squabbling and hostilities among the men (p. 34). A few days later, three quite discouraged sailors abandon the walking party and return to the ship leaving ten men for the trek (pp. 41, 42). Not long after that Albanov begins making very critical observations about his companions because the lack of fuel results in cold meals, which upsets him but not the others: "They [are] incapable of any serious thoughts, but they also lack determination and enterprising spirit ... Serious or critical situations drain them of all their strength (p. 54)." And when they must make unplanned stops due to mistakes and carelessness, he states they "are no better than children (p. 56)." Several days later, Albanov comments on their fading spirits when he declares "my men grow abject and despondent (p. 64)."

The situation further worsens when Albanov discovers that someone has been stealing food and he tells the men that if he catches the person, he will shoot him (p. 89). A few days later two men take off in the night and steal some of the best gear and provisions including boots, warm clothes, biscuits, a shotgun, cartridges, good skis, matches, binoculars, and Albanov's pocket watch, while leaving the others with scant equipment (pp. 97, 98). Meanwhile, there is "no end of squabbling" because Albanov had concluded that dragging their kayaks over certain terrain as opposed to carrying them would compromise their seaworthiness (p. 101). His assessment is that their upset is due to "mental laziness (p. 101)."

He then makes many more comments about their mental state and lack of drive referring to their apathy, dragging their feet, constantly cursing each other, laziness, indifference to his pleas, loss of interest in his plans, and immobility (p. 103). And while conditions such as the depletion of food and supplies seem to only strengthen Albanov's resolve, he declares that his companions "surrendered long ago (p. 105)." And so, it goes. In the remaining pages he offers additional characterizations of the group and subgroups within it including the thieves and a number of individuals who have separated themselves from Albanov. His depictions of the others include being the epitome of indolence and stupid, being of no help, a burden, and useless, lacking character and a sense of duty and responsibility, always sleeping, full of despair, cursing and arguing, and being a sickly group, which is sluggish and very easily disheartened (pp. 115, 116, 122, 123, 130). Consequently, Albanov feels he is "essentially alone (p. 115)."

All told only a few references are made to group RSPs including a very flat burial ritual. All evidence points to infrequent, weak, or at best moderate emotionally charged practices that are of limited rank and impact. The distinct impression one gets is that Albanov's personal program of RSPs involving recurrent diary writing, navigational calculations, and physical exercises are the most consequential activities engaged in by any member of the party. Most commentary unmistakably focuses on the low morale of the group, social discord, a lack of motivation and commitment, deviant behavior, and condemnations of the men by their leader.

As with other expeditions several factors contributed to this situation. The hazardous circumstances faced by the group during their arctic journey were surely significant. Some of these conditions were due to the Captain's flawed decision-making, which resulted in the ship becoming trapped in ice in the first place. Also, the makeup of the crew and their lack of qualifications due to poor crew selection were of crucial importance. Last of all, Albanov's style of leadership may have limited how often RSPs were engaged in by his group. In all fairness, however, it is impossible to know for certain if this was the case.

Albanov was not totally responsible for what occurred. He was thrust into the situation with a crew his didn't pick nor was able to prepare. It should be appreciated that, as David Roberts states in his Introduction to the memoir, Albanov was a very determined, strong leader who struggled against great dangers while successfully navigating the group on their trek. "Without him, the other men would have died early on (p. *xvii*)." At the same time though he was most likely "an autocratic and headstrong leader" who was openly critical of his captain and fellow crewmembers (p. *xviii*). He may not have appreciated the value of rituals and the different needs of the crew and was, therefore, uninterested in or incapable of facilitating practices that might have improved the social dynamics of his group. This was probably even more the case given the disposition of the persons he had to deal with.

The Lady Franklin Bay Expedition: Discord and Loss in the Arctic

Abandoned (1961), an account compiled by A. L. Todd, describes an arctic expedition captained by the American Lieutenant A. W. Greely between 1881 and 1884. The Lady Franklin Bay Expedition, which was sponsored by the United States Army Signal Corps, began with a crew of 25 men who were taken by ship to a bay in the Arctic where they were to establish an outpost, Fort Conger (located in Qikiqtaaluk Region, Canada), for making observations, and conducting scientific investigations. The crew, composed of mostly Americans and at least two Inuit dogsled drivers, began by building a winter house to live in until a resupply ship was to return the following year. As we shall see the expedition began well but dramatically deteriorated as supplies shrank and a schism developed between the crewmembers.

After an initial period of relative calm serious problems arose which severely disrupted the lives of the men. Most important the resupply ships, which were to provide them food, fuel, and other materials, never reached them chiefly due to poor weather conditions involving gales and heavy ice (pp. 50, 64). The failure to reach the outpost in 1882 was "a cruel blow" to the men partly because "There would be no break in the monotony of their confinement, no letters from home [as] they faced another long, dark arctic winter (p. 5)." Then in 1883 the second resupply ship failed to reach them, which meant that, according to a prearranged plan, they had to abandon their camp and make their way by boats to a rendezvous point at another location (pp. 62–64). After more than a month sailing in the icy sea and sometimes becoming frozen in ice, they reached land where they constructed huts to live in.

By this point, however, their food rations began to shrink. Their daily food intake was reduced to about half of what is needed in the Arctic and decreased even further over time (p. 113). By the winter of 1883–1884 the group was suffering from severe privations, which led the Top Sergeant to state in his diary that their typical concern "hinged on the main topic – food" and "The constant gnawing of hunger almost drives us mad (pp. 131, 132)." By now their state of starvation was contributing to their physical decline, e.g., gangrene and one man losing most of his fingers and both feet (pp. 139, 140). Reduced to eating whatever they could find such as dog biscuits, small shrimps, lichen, their sealskin sleeping bags and shoes, bugs, and animal droppings individuals died one by one (pp. 139, 143, 195, 196, 198, 202, 205, 220, 227, 229, 236, 239, 240, 243, 245–248). Finally in June 1884, a rescue party arrived and saved six nearly dead men including Greely.

As bad as it was the mission began on a positive note and Greely saw to it that the crew engaged in various ritual activities. Knowing that the "greatest winter enemy is boredom" he kept his crew busy working and being active in different ways (p. 30). For instance, after establishing their station at the end of August in 1881 Greely, reflecting his Puritan beliefs, dictated a mandatory "psalm-reading" every Sunday morning and minimal duties but no games on that day (p. 24). On a person's Birthday he was relieved from duty, allowed to choose dinner, and given a quart of rum to pass around to his comrades (pp. 31, 38, 39). The men played various games like checkers and cards on a daily basis although no gambling for money was allowed, enjoyed various outdoor activities like snowshoeing, and read from a very large library (p. 32). Thanksgiving was a holiday celebration accompanied by a psalm reading, outdoor sporting competitions, a marksmanship contest with prizes, and a feast (p. 32). A post newspaper, the *Arctic Moon* containing both humorous and serious stories was started but lasted only two months (p. 33). Greely also initiated an evening lecture series but due to a lack of volunteers the Captain had to continue the program alone (p. 39). Finally, Christmas was celebrated with hot eggnog, presents from home, a banquet, and a very entertaining variety show, all of which was followed by New Year's festivities, which included songs, dancing, cigars, and revelry.

Unfortunately, tensions among the officers had also been brewing since the beginning of their mission. In fact, it is around the holiday season that these problems began to fully surface and we find far fewer references to rituals. The next mention of RSPs doesn't come until a full year later when the 1882 Christmas is described as a "pale ghost of the previous one" with few presents and "no enthusiasm (p. 54)." Due to bear tracks around the station, Greely also orders the men to stay close to their house, which prevents them from enjoying long outdoor walks (p. 55). The following winter Greely initiates an evening activity where some men read aloud to each other, and he and others lecture about different topics (pp. 133, 167). And their limited food and rum supplies are pulled together to create an "illusion of a feast" for Thanksgiving and Christmas and a temporary feeling of "euphoria (pp. 140, 141)." Greely also continues "his Sunday reading of scripture" although its effect on the men is unclear (p. 226). The only other mention of rituals involves simple funeral and burial services where the bodies are placed in extremely shallow graves due to the hardness of the icy ground. By the end, individuals do not even bury the body of a comrade or celebrate the summer solstice (pp. 248, 249).

Immediately after the aforementioned holiday celebrations when references to rituals drop off issues involving poor morale increase including problems with certain officers, particular the second-in-command who resigns his post early in the journey because he disagrees with Greely's orders that officers should eat breakfast with the rest of the crew, and the medical doctor who feels he is more educated and intelligent than Greely and should be in command (pp. 41, 42). Greely's fears that the doctor is sowing "seeds of dissension and discontent" seem quite justified given the events to come (p. 45). For instance, the suggestion of a democratic "town-meeting" to discuss what to do to save themselves hints at "mutiny (pp. 58, 59)." As tension rises in the camp, the doctor decides to resign his official position and Greely has to threaten him with a court-martial (pp. 60–62). Later a group of men including the doctor and resigned second-in-command try to persuade the Top Sergeant to mutiny, but he refuses, and the attempt ends (pp. 85–89).

Many additional comments are also made about quarreling, complaints, drunkenness, anger, and thievery among the crew. Several times the engineer becomes extremely drunk and had to be dealt with (pp. 50, 51, 84). Other examples include a crewman who is insolent with the Captain and is demoted, an individual flaring up at the Top Sergeant, Greely demoting another man due to his repeated complaints about the expedition's leadership, and various cases of men arguing, fighting, or becoming upset with each other (pp. 51, 52, 115, 116, 173, 174, 177, 213, 226, 227). Particularly serious are the cases of food theft that take place. A number of instances of stealing by different crewmembers are reported including one with a concealed criminal past and the doctor (pp. 131, 132, 142, 143, 188, 221, 240, 241). Repeated food thefts by the former individual ultimately become so serious Greely has him shot and killed (pp. 242, 243).

Furthermore, various entries in Greely's diary document how poor morale, discontent, conflict, selfishness, and disregard for others permeated the group.

They include Greeley commenting on the bad character of the doctor and the divisiveness of some men, hostilities exploding between Greely and the doctor and another man, Greely writing a critical letter about the doctor's misconduct to be read by a General if Greeley dies, and another fierce dispute between the two men (pp. 219, 220, 222, 223, 228, 237). Finally, problems involving poor morale and social relations are seen in occurrences such as a crewman and the doctor refusing to let the Top Sergeant come into their tent one stormy night which meant he had to sleep outside in the cold and the Sergeant describing one individual as a selfish malcontent who shirks his duties and wants to abandon everyone (pp. 237, 245, 246).

To sum up, this expedition began on a promising note because Greely tried to promote a number of RSPs that appeared to be of a moderately high rank. After approximately five to six months, however, fewer practices took place as problems surfaced within the group and morale declined, followed by no supply ships reaching them. An escalating series of interconnected factors all contributed to the mission's tragic outcome: the grim situation created by dwindling food and supplies, the composition and dispositions of the men which contributed from the very beginning to discord and schisms between officers and different crewmembers, and a deterioration of group relations involving in no small part the decline of effective RSPs.

Concerning the last point, it is difficult to say that Greely provided poor leadership given his being a very serious, dedicated person who promoted many practices in the beginning and attempted with less success to do so later in the mission. The reasons for this may well be due to the fact that some of the officers and other crewmembers were at such odds, which severely undercut Greely's efforts to facilitate group rituals, especially in their bleak situation.

It is noteworthy, as we shall see in other expeditions such as the Scott mission, that where RSPs are an enduring part of the group culture, interpersonal relations and morale can remain strong even when individuals face their likely demise.

The Polaris Expedition to the North Pole: A Recipe for Failure

We now turn to another well-known arctic venture, the *Polaris* expedition, described in Richard Parry's compilation *Trial by Ice* (2001) and Captain George Tyson's memoirs (Blake 1874). This mission (1871–1873), sponsored by the United States Congress, was the first American attempt to reach the North Pole. The original crew of 25 men, composed of 13 Americans, 8 Germans, 1 Russian/German, 1 Swede, 1 Dane, and 1 former British citizen, was later joined by a Greenlandic hunter and 3 Inuits, a male hunter, his wife, and their child. Unfortunately, the expedition never got further than Greenland due to a number of tragic events and problems such as unclear goals for the mission.

Indeed, one could say that the *Polaris* expedition provides a stunning example of what not to do in an undertaking such as this.

According to Parry (2001), the crew faced many perilous conditions, which threatened everyone's lives including freezing temperatures, treacherous ice floes, raging storms, loss of food and fuel, and medical problems such as scurvy. Moreover, after less than three months into the journey their leader, Captain Charles Francis Hall, upon returning from a dog sledding trip and being given a cup of coffee, became violently ill and died within a few days (pp. 107–118). This event created both dismay and suspicions among some of the crew (and Hall himself) due to the mysterious nature of the illness and questionable treatments provided by Dr. Bessel, a German doctor who was in charge of the scientific mission and quite contemptuous of Hall.

In addition to the Captain's unexpected demise, the ship also became trapped in a cove due to heavy ice. The crew then separated during an intense storm, which damaged the vessel. Part of the group was stuck on the *Polaris* until they ultimately abandoned the boat and made it to land. The rest of the crew, who were trying to move supplies and equipment onto the ice during the storm, sought safety in smaller boats, which they then sailed in for 195 days among the dangerous ice floes (p. 260). Rescue of the two groups did not occur until the spring and summer of 1873.

For the first several months under Hall's command, the crew did participate in certain rituals. Work routines involving preparation of the ship for winter and the collecting of scientific material kept "everyone busy (p. 93)." The most often referred to RSP, however, was religious in nature. The fervently religious Hall held daily religious services and special Sunday services where he would give a sermon and urge the men to pray for the ship's safety (pp. 66, 82, 85, 93, 94). With his death all of this came to an abrupt end.

The new captain, Sidney Buddington, almost immediately "canceled the daily services that Hall held" which meant there was no longer a daily collective event for the men (p. 126). Then he announced, "that attending Sunday services were no longer required (p. 131)." For thanksgiving extra food was prepared, however, no special services or celebrations were held (pp. 132, 133). One reference is made to the men in December "playing cards and racing sleds on the refrozen bay" but this is then followed by the observation that there was a loss of order on the ship (p. 133). "Now when the need was greatest to establish regular routines to prevent … malaise … there were none (p. 133)." Then Christmas arrived to be met, as was thanksgiving, with great amounts of food and alcohol, but no religious services (p. 136).

Actually, divisiveness among the crew, which was a normal occurrence, even arose over whether this ritual event should be held on Sunday or the next day (p. 136). At the party presents which were actually items brought as gifts for the Eskimos by Captain Hall were exchanged, but "most were too drunk to care (p. 136)." Other than that, the only other mention of ritual concerns its absence and occurs much later in the mission after the ship is abandoned and Buddington and part of the crew move to land. By this time Christmas and New Years were not celebrated at all (pp. 235, 236). As for the rest of the crew led by George Tyson, the Assistant Navigator, who are either sailing or stuck in ice for months, only one reference is made to ritual which involves their using the last of their "ham and dried apples" to celebrate Christmas (p. 226).

All told the limited range of rituals during the first three months that mainly involved religious services and work were drastically curtailed after that. Very few comments are made about RSPs and most of those mentioned such as Christmas involve events that are impoverished, poorly performed, a source of conflict, and/or not practiced at all.

While divisiveness and problematic group dynamics are certainly evident in the first months of the expedition their presence increases over time. This is due to a number of factors in addition to the loss of RSPs that haunted the expedition from its beginning to its end. To begin with a unique authority structure existed with leadership divided between three captains (p. 33). Hall was the overall commander and Buddington and Tyson were the other two captains although what this actually meant, especially for Tyson, was not totally clear. Moreover, while this was an American expedition a significant number of the crew was German, which led to divisions "along the lines of nationality (p. 40)." Then, even though this was a Naval mission the sailors were not military but civilians who had previously worked on whaling ships. And another group involved several German scientists including Doctor Bessel who are described as extremely "haughty (p. 43)." Actually, the top priority of the mission itself was not clear, i.e., the goal being "exploration or science (p. 43)."

All of this was extremely important because very quickly quarrels and disagreements arose between Hall, some officers, and the scientists over who should command. These fights then spread to the crew as a whole who divided along "national lines (p. 51)." Adding to the mix "three divergent groups" travelled on the *Polaris* (p. 76). Buddington and others including the English-speaking sailors were mainly concerned with the safety of the ship. The scientific corps and German crew wanted to study the region from the *Polaris* or territory close by. In contrast Hall, a few other sailors, Tyson, and the Inuit Eskimos wanted to leave the ship and explore the land. Hall who had Arctic experiences of this sort was determined to travel by land and reach the North Pole, which "threatened the safety" of the other groups (p. 76).

The character and experiences of certain authority figures are also important. Captain Hall had experience leading small groups but not large parties (p. 42). He also "micromanage[d]" people's work while sometimes ignoring other basic needs leading to a resentment among some over his "imperialism" and the belief he was "in over his head (p. 99)." Buddington had a serious drinking problem. And Bessel was extremely arrogant, rigid, and held exaggerated ideas about his abilities and importance.

Problems among the crew both before and especially after Hall's death were manifested in many ways in addition to those already mentioned. For instance, a special boiler for seal oil or blubber was sabotaged while Hall was in command and later Hall tells Tyson he has left important papers ashore for safekeeping out of what seems to be a sense of foreboding for the expedition or possibly himself (pp. 61, 77, 78). Not long after his death we find that "stability aboard the *Polaris*" began to almost totally vanish as Captain Buddington and the men were increasingly drunk (p. 133). Instead of day-to-day RSPs which could help prevent malaise "There is no stated time for putting out lights; the men are allowed to do as they please," which includes nightly "boisterous, drunken parties (pp. 133, 134)." A lack of discipline had become the norm for most of the crew and Captain Buddington who "remained drunk almost daily (p. 139, 144)."

Other related issues also affected the crew. For instance, Buddington and Bessel disrespected and held great animosity for each other, which led to arguments (pp. 136, 137). Partly due to personal dislikes, if not hatred among both officers and the enlisted men, a "splintering of loyalties" occurred and "anger added to anger" all of which was reflected in the "invectives" which filled the diaries of some of the crew (p. 137). Reflecting the lack of discipline of the group and the poor leadership of the new captain, the crew actually brewed beer instead of making needed repairs on the *Polaris* (p. 154). At the same time "a growing, open contempt for him developed on the part of his crew" as witnessed in the refusal of crewmembers to obey the orders of Buddington to hand pump water from their ship (p. 168). In essence "discipline and cohesion ... had ... dissolved by degrees over the long winter [while stranded in ice]" until "the chain of command had virtually vanished from the crew" leaving "an unruly, self-serving mob bent on having their own way with no regard for the consequences (p. 169)."

This situation contributed to a number of bad decisions two of which stand out. While trapped in the ice the decision is made to set off in a completely unrealistic and extremely dangerous 600-mile trip to reach the North Pole in two "fragile whaleboats (p. 155)." "Naivete, lack of experience, and alcohol" were at least partly responsible for this attempt, which ended in a few days after one of the boats was crushed by ice (p. 155). The other decision came after the crews were separated. The search for the part of the crew in small boats among the ice lasted only one day and was then discontinued. This was most certainly due to indifference and a lack on dedication to their shipmates, which was fostered by group divisions.

Conditions were not any better for the other group made up of ten crewmen, the Eskimos, and Tyson who were trapped on the ice in smaller boats and huts they sometimes constructed. The sailors were very undisciplined and made foolish decisions such as using one of their whaleboats for firewood, but because Tyson "had never formally commanded the crew, and ... had no firearms while they did," he could not stop them (pp. 218, 219). Moreover, friction between Tyson and another German officer, Meyer, the navigator, never stopped (pp. 219, 220). As Parry states: "Instead of working together, they fought ... the officers of the *Polaris* still engaged in a struggle for control ... (p. 247)." In general, these divisions spread through the entire group pitting "the Germans who camped sullenly in their igloo" versus "the Inuit, ... an Englishman, and ... the black cook [who], sided with Tyson (p. 246)."

So too as food supplies decreased "stealing and hoarding rose (p. 220)." The theft of foods such as bread, canned meat, pemmican, and chocolate, became an increasing problem (pp. 20, 222, 228, 229, 256). And hardship and starvation "only widened the gulf between the various factions" and the fighting between them whether that involved the officers, German and non-German crewmembers (e.g., since they reverted to speaking only German in Tyson's hut, he moved in with the Eskimos who spoke English), or other factions (pp. 222, 223). For instance, after one of the Eskimos named Ebierbing shot a seal "the crew snatched the animal away from the hunter and dragged it into their igloo" where it was eaten in minutes (p. 246). Not surprisingly the "epidemic of criticism" between the different factions came to include complaints about the sailors by even the Eskimos who were usually not that vocal (p. 248).

In essence, the *Polaris* expedition was marked by discord among both officers and crewmembers, different factions, a loss of cohesiveness, contemptuous relationships, a lack of discipline and order, low morale, thievery, and murder. We say murder because, in 1968 a four-man team flew to the gravesite of Captain Hall and returned with a tissue sample from his frozen body to be used in an autopsy. Neutron-activation analysis conclusively showed that Hall had been poisoned with arsenic (p. 301). While we cannot know for sure who was responsible, Parry who is a surgeon concludes his examination of the evidence with the assessment that Bessel "is the most logical choice" and Buddington may have been an accomplice (pp. 304, 305).

In sum, contributing to the problems experienced by the crew are a number of factors involving (a) the extreme conditions they faced, (b) the composition of the crew which included very diverse subgroups, (c) their dispositions, backgrounds, apparent prejudices, and training or lack thereof, (d) the organizational structure of multiple and sometimes opposing captains and officers, and last but not least (e) limited RSPs which deteriorated or were terminated.

Concerning the latter, leaders exhibited varying degrees of effectiveness for implementing RSPs. Hall demonstrated some interest in promoting work and religious rituals, which reflected his own personal preferences. These probably had at best a low to moderate rank and impact on the crew. Buddington appeared to be totally disinterested in, if not opposed to, any type of rituals on the ship. And Tyson may have wished for certain types of ritual practices but faced a quite difficult, if not impossible situation due to the opposition to him by others on the ice floes. With that said our second account comes from Tyson himself.

Arctic Experiences, edited by Vale Blake (1874), is Captain George Tyson's personal memoirs about the *Polaris* expedition. Tyson's own account provides a quite informative companion piece to Parry's compilation. Having already described the mission we will focus on Tyson's discussion of events while still on the ship and then on the ice.

In the first three months of the journey before Hall's death the only references Tyson makes to rituals and collective events concern Captain Hall's religious services. Tyson refers several times to these services and prayers led by Hall (pp. 150, 151, 153, 154, 159, 160). After Hall's death, burial ceremony, and Buddington becoming the commander Tyson then reports, as does Parry, that Sunday services were stopped and daily service was no longer compulsory (pp. 162, 165, 166, 168). He also says that Thanksgiving was remembered at the dinner table "but in no other way" and there was "No service on Christmas (pp. 169, 171)." After these allusions to dead or non-existent rituals, he reiterates that Sunday service was not held and once when it was most of the crew did not attend (pp. 175, 176).

Concerning morale and group dynamics Tyson states, "there is not perfect harmony between Captain Hall and the Scientific Corp, nor with some others (p. 142)." A couple months later in December after Hall's death Tyson then makes the troubling observation that "Nothing occurring that is pleasant or profitable to record. I wish I could blot out of my memory some things which I see and hear (p. 171)." He then declares, "Captain Hall did not always act with the clearest judgment, but *it was heaven to this* [Tyson's italics]. I have not had a sound night's sleep since the 11th of November (p. 171)."

After being separated from the ship and stuck on the ice floes for over six months Tyson makes several more comments, often of a negative nature, about rituals. He states that the Thanksgiving breakfast and dinner included a small "extra allowance" of food from their meager supplies but "there was no general religious service," which made it a relatively empty celebration (pp. 222–224). His most positive comment concerns their "Christmas feast" where they had extra food even though the amount eaten was very small by normal standards (p. 233). A few references are made to simple, personal rituals Tyson engaged in involving stargazing and smoking his pipe until he runs out of tobacco, which he says is "his only companion (pp. 247, 251, 283, 293)." He also laments not being able to read on a regular basis due to the lack of reading material (pp. 257, 258). The only other group rituals he mentions are games of checkers, chess, and cards played by his Eskimo companions (pp. 261, 262, 271). Finally, Tyson notes in his diary how "festivities" were occurring back in the United States for Washington's birthday but he doesn't have the "spirits" to raise a flag because "no one here ... knows or cares any thing about Washington – foreigners all (p. 287)." After their rescue Tyson also stresses the importance of certain collective RSPs when he attends a religious service on a ship, the *Tigress*, and states how pleasurable it was after being "deprived of united religious services for sixteen or eighteen months (p. 334)."

The crew on rare occasions is in good spirits as when they become extremely happy about a large seal kill and having plentiful food for a period of time (p. 291). However, overall, Tyson paints a very negative picture of the group. He criticizes the lack of "discipline" of the men with him on the ice and on the *Polaris* and speaks of "annoyances" such as the language barrier because many of the crew speak German all of the time, which he doesn't (p. 221). He also complains about the total breakdown of authority as was the case on the *Polaris* referring to the "pilfering" that occurs much of the time and how he "can scarcely get an order obeyed (p. 226)." In that regard, when their lives depend on struggling to prevent their boat from sinking in very rough water and he orders them to give their all, he says "for once" they obeyed (pp. 321, 322). More dramatically he even mentions the concern of one of the Eskimos and himself that some of the crew might kill the Eskimos and commit cannibalism (p. 230).

The state of mind of the group is quite evident in statements about how he has rarely seen a smile on anyone's face and "how strange it would sound to hear a good hearty laugh; but I think there never was a party so destitute of every element of merriment as this (p. 250)." And concerning the "animus of some of the men" Tyson describes how he was verbally "abused" and threatened with "personal violence" in an unprovoked manner by one of the men who entered his tent (p. 250). He then expresses the fear that "this business will end ... unless there is some change ... in a disastrous manner (p. 251)." His overall opinion concerning the disorder and low morale of himself, and the crew are reflected in his assessment of the situation:

> Not a countryman of my own on the ice - all foreigners. Not one to talk to or counsel with ... with an utterly undisciplined set of men; impossible to get an order obeyed or to have any thing preserved which it is possible for them to destroy. They take and do as they please. (p. 278)

In essence, Tyson's memoirs are highly compatible with and corroborate Parry's account of the expedition. They are consistent in their descriptions of how RSPs are limited in nature, often weak, empty, and sometimes terminated. And they are consistent in their depictions of the problems plaguing the crew, which includes poor morale, divisive factions, a lack of communication, extremely troubled relations, and ineffective leadership due to various reasons.

The Belgica Expedition in Antarctica: Amundsen's Account of a Flawed Mission

Our next mission, conducted in 1897–1899 was the first scientific expedition to the South Pole and to winter in Antarctica. Subsidized by the Belgian government the crew of 25 men was composed of several nationalities: twelve Belgians, eight Norwegians, two Poles, one American, one French, and one Romanian.

The Belgian expedition is described in *Roald Amundsen's Belgica Diary*, edited by Hugo Decleir (1999). While Amundsen, who served as a second mate on the expedition and later became a very renowned polar explorer, stays in relatively high spirits what the crew experiences is of a very different nature.

Not surprisingly, the crew faced many difficult conditions that disrupted their lives including their ship, the *Belgica*, becoming trapped in ice their first Antarctic winter. Other problems encountered when they were sailing, land exploring, or trapped in ice included storms, no sunshine for extended periods of time, snow blindness, not enough food, not having the proper tools, their tents being too small and fragile, and their sleeping mattresses becoming wet and frozen on the ship (pp. 77, 79, 136, 173, fn 2, 178, 97, 99, 120). Even worse, however, were the physical problems experienced by some crewmembers and the death of several men (pp. 132, 175). For example, one man fell overboard and drowned while another died of heart problems (pp. 70, 71, 110, 111). And some persons experienced mental problems.

Various references are made to rituals engaged in by the men, however evidence reveals that most of the comments pertain to a small subgroup and that the RSPs for the crew as a whole are more limited in number and significance. Rituals involving the crew include celebrating their first Christmas early in the journey, commemorating the Norwegian national holiday at a mid-day meal, the Belgian commander, Adrien de Gerlache, celebrating the American holiday (July 4) with a simple toast for their physician, Dr. Cook, celebrating the Belgian National Holiday with a three day break, celebrating the one year anniversary of the expedition's beginning with champagne, a toast, and the crew having the afternoon off, their being allowed to ski to an iceberg where Amundsen organized a jumping competition for the men, and many of the crew going to the iceberg a few days later for their enjoyment (pp. 58, 103, 116, 121, 134, 137–139). Other than that, the only other mention of RSPs during the two-year mission involve celebrating New Year's Eve (1899) with a party and cognac supplied by Amundsen although not everyone attended including the captain, a spontaneous celebration of the first appearance of the midnight sun, and a party close to the end of the journey in which music is played (pp. 170, 171, 173, fn. 2, 173, 174, fn. 9, 197).

Most striking, however, are the far greater number of references to Amundsen's personal activities which he regularly engages in both alone and with others, particularly Dr. Cook, Lecointe, the first officer, and occasionally Arctowski, the meteorologist and geologist. Aside from a few remarks about activities such as playing the card game whist (p. 128), nearly all of the commentary concerns Amundsen and sometimes the other three men repeatedly skiing, going on walks sometimes using snowshoes, and occasionally taking photographs. At least 26 references, sometimes involving multiple comments on the same page about separate RSPs, are made to these activities, a number which is approximately two and a half times more than references to RSPs among the entire crew (pp. 90, 91, 95, 99, 109, 112, 113, 117–121, 125, 126, 137, 144, 145, 146, 149, 161, 172). The importance of these highly salient, oft repeated outdoor practices cannot be overstated as recognized by Decleir, the editor, who points out how they provided "recreation" and an "active reaction" to the environment, which protected Amundsen from depression (pp. 127, 125). Furthermore, while quite different in nature, the RSP of keeping a detailed diary was also highly beneficial to Amundsen's state of mind.

At the same time, numerous observations are made about the absence, weakness, or fractured nature of RSPs among the crew. Many comments highlight the lack of group rituals such as religious services or celebrations on special occasions, e.g., Maundy Thursday, Good Friday, Easter, Sunday mornings, Amundsen's birthday, or the appearance of the sun after a long period of darkness (pp. 93, 97, 119, 120, 136). So too the editor points out that 'Midwinter" (winter solstice) which is "the most important day to be celebrated during the winter in the Antarctic" elicited no response from the crew (p. 122, fn 14). A day off is given because of All Saints Day, a significant holiday in Belgium, but no collective activity is noted other than Amundsen going on one of his ski trips with Lecointe and Cook (p. 149). And Christmas Eve passes "quietly" with the men receiving cigars and clothing but otherwise "The evening was spent like every other evening (p. 169)." Clearly a weak ritual as noted by the editor who states: "It was obvious at Christmas that there was no longer a desire to hold parties (p. 174, fn 9)." He also says that on New Year's Eve, "according to Arctowski, it looked as if no one was in the mood to celebrate" and several crewmembers had already gone to bed before Amundsen, as previously noted, broke out several bottles of cognac and was able to start a party with some of the men (p. 174, fn 9). It is also striking that when the expedition officially ends at Easter time no mention is made of any type of celebration.

Lastly, differences between two ritual events are symbolic of a basic split and deep-seated nationalistic biases among the crew. In recognition of Norwegian Independence Day champagne was provided at a meal and the crew were given the afternoon off (pp. 102, 103). Two months later when the Belgian National holiday arrives it is celebrated in a much grander fashion with the men getting "three days holiday," the raising of their flag, the drinking of champagne, and music (p. 121).

Aside from one statement about how most of the crew gets along with each other, numerous comments document how they experienced low morale, a lack of cohesion, and even psychological problems (p. 93). For instance, the editor discusses "a lack of discipline" in the earliest stages of the expedition along with the fact that the death of a crewmember saddened the crew (pp. 32, 71).

More precisely several comments describe conflict, arguments, threats, and thievery among crewmembers. For example, at one point in the expedition all the officers and scientists disagree with the commander's decision to trek south because of illness among the crew and inadequate equipment (p. 125).

The editor also reports that a number of disagreements arose involving the commander versus Amundsen and/or Lecointe at different stages of their journey (pp. 154, 175). And a crewman threatens to murder another man while four men are discovered to have stolen many provisions. In this regard, Amundsen expresses relief that they are not Norwegian due to the prejudice some have for his countrymen (pp. 125, 148). Amundsen also becomes upset due to his learning about a contract that officially eliminated his authority on the ship. He feels tricked but continues to work out of a sense of duty (pp. 154–156). Quite pointedly Dr. Cook declares the contract "had drawn a line between the honest Belgians and the dishonest foreigners (p. 155)." After that Amundsen's tone in his descriptions of the Commander and other Belgians becomes more negative (p. 173, fn 3).

What's more numerous references are made to the mental health of different persons. Several crewmen at different stages of the expedition are described as "out of their mind," deranged, or insane (pp. 132, 150, 153, 161, 167, 175). Other commentary by the editor based on several sources discusses how nearly all of the crew, both when the ship is stuck in ice for months during the polar night and later in the spring when the sun returned, suffer from "despondency ... and mental depression," a "general lethargy ... loneliness and hopelessness" and low "morale (pp. 123–125, 153)." He also says that they had lost a "desire to hold parties" and "the unchanging situation outside the ship seriously affected the general state of mind (p. 174)." He then stresses that even "The summer months [were] ... a great disappointment" because "The general state of health did not improve spectacularly" for the crew including the Commander who "drew up his will in the event that he ... were to die" (p. 175). And both Amundsen and the editor point out at different times that the Commander's and others "mental state" is not good and a cause for worry (pp. 136, 153).

Actually, Amundsen makes one critical comment about the expedition's leadership, which helps to explain why this situation exists. He states: "We could all do with some sun and air but unfortunately we do not have leaders who are intelligent enough to realize this. So, the men spend these fine days ... in the musty, hot air of the ship, working on things which are not in the least urgent (pp. 139, 140)." He recognizes how important it is that they have a change from their work and be outdoors. And we would add to engage in meaningful RSPs.

Finally, prejudiced beliefs involving Belgians and Norwegians are conspicuously present in this mission. In addition to several previously discussed developments such as Amundsen's contract and loss of authority, he squarely addresses the issue when he says, "The commander always talks about the inbred slowness of Norwegians" who are held in "little esteem (p. 162)." He then goes on to state: "There are so many differences in thought and emotions between us and the Belgian nation that it will never be possible to collaborate really well together (p. 162)." Such cultural and nationalistic biases were a source of tension, frustration, uncooperativeness, and weakened solidarity among crewmembers.

In sum, the *Belgica* expedition was social psychologically speaking a flawed mission. These difficulties were a product of several factors including the challenging conditions they faced. At the same time evidence indicates the commander's decision-making contributed to some of the disagreements between himself and others. And while not as extreme as what occurred on the *Polaris* the split between different nationalities grounded in prejudiced beliefs impeded the development of shared ritualized practices among the crew, undermined the cohesion and state of mind of the group, and fostered disputes between the commander and others. Furthermore, the unchanging situation they were in for extended periods of time contributed to depression, lethargy, and various mental problems among the men. The creation of meaningful, salient, repeated RSPs that could have provided a counterbalance to these conditions would have been extremely helpful. But apparently the leader was focused on the tasks and duties carried out by the crew and not the social and ritual dynamics of the group. The benefit of such practices is clearly evident when we look at the recurrent RSPs Amundsen and several others engaged in and their higher morale.

The HMS Enterprise in Search of the John Franklin: Conflicts of Authority and Ritual Failure

Our last failed expedition is quite different from those we have already examined. The *HMS Enterprise*, a British Arctic expedition (1850–1855) led by Richard Collinson, was commissioned by the Admiralty to search for the famed, missing John Franklin expedition. The crew of the two-ship mission – the HMS Enterprise and the Investigator – was composed of 59 men, which available evidence indicates was all British. William Barr (2007) in *Arctic Hellship* provides us with a careful, well-researched, and balanced description of the expedition and many of the social forces at work on the ship.

The crew certainly faced disruptive, challenging conditions similar to those encountered on other expeditions. For instance, extremely wet weather made many people sick, several crewmen died due to health problems, and one man committed suicide, which many believed resulted from the captain's insufferable treatment of him (pp. 59, 82, 83, 131, 175, 232). Certainly, the winters, as in other missions, were difficult because they could not sail the ship and had to live on the immobile vessel. And the duration of the voyage, over four years, undoubtedly made this a very trying experience.

As for ritual practices engaged in by the men our examination of the evidence reveals a major difference between the two most important groups on the vessel: the executive officers and the rest of the crew. We find almost no explicit commentary about rituals engaged in by the officers. Essentially only a couple comments are made to RSPs concerning the executive officers,

Warrant Officers, and Ice Mates who enjoyed a "splendid Christmas dinner" one year and the officers who periodically played with dice (pp. 156, 116). When playing dice they would gamble with very small amounts of money, however, the captain "launched an attack on all of his officers" and forbid betting, an activity, which could be considered a recreational RSP (pp. 115, 116). While the rest of the crew did participate in RSPs no mention is made about the officers being present or enjoying these activities. The only other reference to ritual concerns some of the officers who fairly early in the voyage stopped attending the weekly religious services held by Captain Collinson (p. 110).

Concerning the crew an officer on one occasion tries to get some of the men to sing but they wouldn't because they were sulking over their bad tea. Later in the expedition certain rituals also lost some of their appeal since they had been engaged in during previous winters (pp. 91, 201). Nevertheless, the crew participated in a number of practices, which they appeared to enjoy, many of them occurring during the winter months. As already noted, the captain carried out regular church services for the men (pp. 110, 113, 140, 208). Various activities are also referred to such as celebrating Guy Fawkes Day, playing rounders (a British version of baseball), cards, cricket, and having the crew go for a walk on Thursday mornings (pp. 115, 153, 202, 204). Christmas celebrations were also held involving church services, large meals, and sometimes drink, dancing, and singing (pp. 115, 156, 230).

The captain also initiated a number of other activities. A school for the crew was established during the winters in which academic subjects, tailoring, and shoe making were taught in the mornings and afternoons (pp. 114, 157). A high-quality billiard table was built and was quite popular with all of the crewmen (pp. 155, 156, 201, 204). He also had skittle alleys, a British bowling game involving nine wooden pins, built during the winter months, which the crew could play (pp. 154, 204, 210). The most referred to activity, however, involved theatrical performances that crewmembers would prepare and then usually present every one or two weeks (pp. 115, 154, 155, 157, 158, 159, 160, 201, 203, 222). Quite a bit of work went into these productions that were well received by the men.

Barr's conclusion is that the captain "tackled the problem [of keeping the men busy in winter] with a great deal of imagination (pp. 114, 115)." He established a program of recreation and exercise to keep them occupied. Perhaps not surprisingly we find that "Collinson was very satisfied with the effect that the theatrical performances had had on morale" along with all of the other activities the men participated in (p. 160). All told evidence suggests that the captain treated the crew well, was supportive of them, and promoted enjoyable RSPs that were salient and frequently practiced by the men.

On the other hand, Collinson and his officers were in conflict over routes, mission parameters, and other matters from the beginning until the very end of the expedition. The officers saw Collinson as woefully inexperienced in Arctic conditions, which he was, and Collinson appeared to view any input from his officers as personal attacks. In point of fact references to conflict, poor morale, tensions, and the breakdown of authority involving the captain and officers are so numerous we cannot discuss all of them here. We will, therefore, provide several examples of some of the problems they experienced.

In the beginning of the journey signs of trouble are already evident as seen in the journal entries of the master executive officer, Skead, who raises serious concerns about the circuitous course Collinson has chosen for their journey (p. 26). Shortly after that the captain responds to Skead's suggestions for how close to sail near the coast in an authoritarian, negative manner (p. 31). More disputes over the route they are taking follow between Collinson and all the officers (p. 35). Shortly afterwards the captain responds to Skead's route suggestions as "damned foolish," clearly revealing the lines of contention between the captain and his officers (p. 37). Subsequently disagreements and tensions only mount over the route being taken and the captain's premature, from the officers' point of view, decision to "winter in Hong Kong, rather than staying as far north as possible, with a view to getting an early start on their search for the missing Franklin expedition in the following navigation season (p. 49; see also pp. 45, 47)."

Soon after Collinson reports in a written communication his officers for neglect of duty over a seemingly minor issue and restricts time allowed for smoking, which leads to excessive complaints and his rescinding the order (pp. 59–62). Meanwhile Skead's upset grows over the captain's "almost paranoid excessive caution" in his sailing of the ship (p. 62). The feud between the two boils over with a clash of words resulting in Skead being stripped of several duties and threatened with being discharged (pp. 73–78). Not surprisingly Skead, who is the navigator, feels he is being harassed while Collinson incorrectly alters the navigation of the ship misusing maps and instruments (pp. 80-81). As other officers begin to feel vulnerable to the disapproval of the captain, he starts treating Skead with open contempt (pp. 85, 87). At this point several references are made to Collinson being drunk and his exhibiting poor leadership through mistakes and contradictory orders, an accusation made by all the officers (pp. 94, 96, 102).

Skead's outrage with the captain continues to grow when he decides to winter the ship five weeks before ice, wasting search time (p. 111). Then Collinson places Skead under arrest after he converses with other officers about his frustration with the captain, a severe punishment for, as Barr puts it, "a 'trivial offence' (pp. 117–121, 124)." Shortly after that Collinson reprimands another officer for poor performance (p. 126). By now the author states the captain is "clearly out of control (p. 138)." Apparently, this is the case since he next suspends another officer from duty with the intent to court martial him for relatively minor offenses (p. 147). Without a doubt, tensions on the ship were "growing steadily (p. 152)." The captain charges Skead, who has been arrested and confined to his quarters, for another vaguely described offense (p. 172). After a conflict between two officers, Collinson then dismisses both from duty, eventually reinstating one (pp. 178–179). With only one officer left on duty conditions further deteriorate which has "a very negative effect on the atmosphere and discipline (p. 182)." Yet, Collinson remains in total "denial about the problem originating in any way with himself," squarely blaming the officers for the "problems on board (p. 209)."

After increasing restrictions and isolation of Skead for allegedly uttering an offensive phrase from his cell, the captain then decommissions and arrests the last officer for an unknown offence (pp. 219, 220). Now all four executive officers are under arrest with no remaining officers to help run the ship. Finally, during their return in the final year Collinson suspends the leave of the Assistant Surgeon for the alleged bad behavior of calling a troublesome crewmember a "bloody fool (pp. 225, 226)." When the physician protests the treatment of the officers, the captain charges him with more offenses (pp. 225–228).

In a nutshell, Collinson and the officers did not engage in any meaningful, collective RSPs. And morale and social conditions were to put it quite simply awful. The captain exhibited very poor leadership and relations between he and the officers were volatile, marked by disagreements, tension, a breakdown of authority, a lack of communication and cooperation, and little to no social unity. We should also add that the officers themselves did not always engage in the most constructive behaviors, e. g., tensions existed among some of them and the crew.

In contrast, the crew had better morale and social relations due to the captain's more positive attitude toward crewmembers and his efforts to promote RSPs among them. Barr's assessment of the situation concisely describes how "striking is the contrast between Collinson's treatment of his men and of his officers. Towards his men he was … quite benign and solicitous … And … his efforts to provide entertainment for his men … are in striking contrast to the regulations imposed by other Royal Navy captains of wintering ships (p. 243)." Furthermore he "was one of the most ardent supporters of the theatricals staged by the men … [he] even lent items of clothing from his own wardrobe and made his cabin available as the dressing-room for the men (p. 243)."

All of this stands in marked contrast with "his more draconian orders directed at his officers … and his unparalleled behavior in placing all his executive officers under arrest for what can be characterized objectively as very minor (even insignificant) infractions of regulations (p. 243)." By the end of the expedition all of the executive officers were arrested, some for several years while on the ship, decommissioned, and formally charged, i.e., court martialed. Fortunately for the officers all charges brought against them were dropped after they returned to England.

We should appreciate that in regard to the crew's morale and the captain they were quite fortunate to have a number of men with adequate, if not excellent, acting skills (pp. 158, 159). There was also at least one high-spirited crewmember who had an uplifting effect on the men. After his death he is described by another shipmate as "the life and soul of the whole ship's company while on board, always joking and skylarking … and also the principal play actor on board during our long winter evenings (p. 232)."

At the same time, several executive officers adversely influenced the morale of different individuals. One lower ranked officer writes about a "treacherous accusation" made by two executive officers against him (pp. 202, 203). Later he refers to another accusation by one of these officers and complains that he is "always prying into everybody's business (pp. 229, 230)." And the ice-mate also experiences problems with this same officer, which leads the ice-mate to ask to be relieved of duty on the homeward journey, which the captain does (pp. 229, 230).

Over the duration of their long voyage, it is also extremely likely that crew-members, despite their shared RSPs, were negatively affected by the poor relations and bad feelings between Collinson and the top officers. Barr states that: "the attitude of the men (or at least some of them) towards the officers had by this time been seriously poisoned, and that normal naval discipline had been completely undermined (p. 226)." Later he refers to the negative feelings between the ice-mate and the officer, and the latter's formal complaints about the former, as "symptomatic of the vicious undercurrents swirling around the ship (p. 230)."

In essence this was a failed venture, not simply because it failed to achieve the goal of discovering what happened to the Franklin expedition, along with the fact that almost all the coasts they covered had been explored by others, but because of the social psychological dynamics on board the ship, especially those between the captain and top officers and ultimately at least part of the crew (pp. 234, 235). Ironically the captain was sensitive to the needs of the crew and facilitated various RSPs among the men. But impoverished ritual dynamics, divisiveness, and animosity among the expedition's leaders ultimately jeopardized the entire mission.

Conclusion

A variety of causes ranging from lack of experience, inadequate training, poor judgment, and the temperament of expedition members to the composition of the crews, the structure of authority, ineffective leadership, and social prejudices contributed to the absence or weakened state of rituals and the failure of the missions just examined.

Of course, we should remember that these are complex issues as seen in the last failed mission where the captain promoted certain rituals among the crew but not the officers or in other expeditions where the author maintained a personal program of RSPs. In point of fact RSPs such as these provide a good segue to our next two chapters where various kinds of rituals are more pronounced and consequential for most if not all the members of the expeditions we shall be examining.

4

SUCCESSFUL EXPEDITIONS

J. David Knottnerus, Kevin Johnson,
and James D. Mason

As already emphasized, all failed, successful, and extremely successful expeditions encountered problematic conditions that contributed to deritualization among crewmembers. These hazardous ventures created a disruption in the daily rituals of people, which shaped their normal lives apart from their expeditions. However, whether individuals on these polar missions engaged in reritualization is another matter. In this chapter, we turn out attention to endeavors where RSPs are stronger and typically occur to a greater degree.

We will examine six expeditions that fall into the successful category. These missions, like the ones just discussed, differ in a number of ways including what their goals were, their size, when they occurred, and the make-up of the crews.

The Voyage of the Isabel in Pursuit of the John Franklin: A Successful Journey to Greenland

Our first expedition like many others went in search of the lost Franklin mission. The description of the journey comes from the account written by E. A. Inglefield (1853), captain of the British ship, *Isabel* in *A Summer Search for Sir John Franklin with a Peep into the Polar Basin*. Inglefield describes one of the shortest expeditions examined here. Departing from England in July 1852 they returned only four months later in November 1852. Privately funded by Lady Franklin, the mission that was composed of 17 British crewmen in addition to the captain took place in the Arctic region primarily around Greenland.

As we shall see, successful expeditions such as this one, often seem less dramatic in a social psychological sense than failed missions because the overall mood of crewmembers is better and social relations are more harmonious.

DOI: 10.4324/b23044-4

The pathos and angst found on many failed expeditions are less present on more successful ventures. This does not mean, however, that the disruptions they experience are any less threatening.

Even a short four-month voyage such as this one, encountered numerous dangers. For nearly the entire trip the crew of the *Isabel* was subjected to appalling weather and hazards while sailing in the polar seas. Inglefield describes repeated instances of squalls, gales, high winds, cold temperatures, snow, large waves, and fog. Sometimes these gales which could last for days at a time forced the crew to postpone activities such as "divine services," flying a flag on an island, or even heating tea because water from a storm would put out the fire needed to heat food and drinks (pp. 12, 72, 100). The ship also encountered on numerous occasions drifting pieces of ice and icebergs, which placed the crew at serious risk. Fortunately, they were able to navigate their way through these obstacles without any major damage to the vessel or loss of life. All of these conditions are discussed throughout the captain's memoir (pp. 7, 8, 12, 40, 72–74, 79, 81, 89, 90, 91, 94–96, 99–101, 110, 111).

Concerning the behaviors and state of mind of the crew absolutely no mention is made in Inglefield's account about rituals being rarely or never engaged in, their being weak or diminished in any way, or of any conflict between crewmembers, including the officers. Nor are any reports made about poor spirits among the men, weakened cohesion, or troubled group interaction. Rather, a number of quite explicit references are provided about RSPs the crew participated in and the captain's desire to foster good morale among all members of the group.

Several important statements by the captain in the first pages of his account highlight these points. For instance, he says: "I resolved to have nothing different from my crew, no servant, and my provisions the same, and served at the same hours as theirs; by these means I hoped to prevent the possibility of anything like discontent, should hardships or privations be our lot (p. 1)." In other words, his first comment concerns the crew and maintaining their good spirits. And nothing in his memoir suggests that he failed to achieve this goal.

A few pages later he provides his first comments about ritual practices when he says:

> I gave out to the people the Bibles and Prayer Books, which had been generously supplied by the Society for the Promotion for Christian Knowledge. Divine service was performed in the forenoon, and it was most satisfactory to see the attention of all the crew to this important duty. (p. 6)

The next day he informs us that "Being Sunday, the crew were mustered, the lower deck inspected, and divine service performed, after which I read to the men one of those excellent discourses written for the use of seamen by the Rev. Samuel Maddock (pp. 6, 7).'' From the very beginning of the voyage, the captain formally promoted RSPs involving religious services and the reading of religious materials.

These highly prominent and repeated practices stand out as the most dominant ritual activities on board the ship. Additional references to these activities and expressions of religious beliefs and thankfulness, and how they were sometimes interrupted by bad weather or difficult sailing conditions, are found in different places in his memoir (pp. 12, 13, 22, 26, 75, 103, 112–114). While he does not literally refer to every time they occurred, it is quite clear these RSPs were a normal part of the men's lives as reflected in comments concerning "how we met together for our usual divine service" and a "special thanksgiving" was offered after they escaped from hazardous ice floes that were threatening their ship (p. 75).'' Moreover, the services were quite popular. In the beginning of the voyage, Inglefield invited officers and crew to his cabin for prayers but so many men came he then initiated a "regular Sunday evening service on the lower deck (p. 13).''

These were not, however, the only RSPs people participated in. Different comments are made about various work rituals including scientific activities usually involving Inglefield or the doctor. For instance, the doctor used a towing-net to catch and examine seaweed, shellfish, mosses, and various marine animals (pp. 11, 12, 26, 58, 59). The captain also engaged on a regular basis in different activities such as making sketches of lands they sailed by, taking many observations and measurements with his sextant to accurately locate the coastline, and collecting rock specimens when on land (pp. 26, 33, 38, 66, 67, 90, 105).

A number of collective ritual events, shared meals, and visits, especially with officials and indigenous people living in or near Danish settlements, were also enjoyed by different members of the crew. When visiting the Danish Governor at a settlement, the captain and doctor were invited to share an excellent meal with the likeable Governor's family while milk and codfish were sent to the ship for the rest of the crew (pp. 18, 20). The next day, the Governor and his wife visited the ship where the Danish flag was flown and saluted with the firing of guns, gifts were given to the children and Governor, and they attended a religious service (p. 22). Wine and biscuits were then provided to the guests followed by the captain returning to shore to enjoy another delicious dinner (p. 23). Inglefield then attended a Christian religious service in a "native church" conducted by the Esquimaux (pp. 23, 24). The service included the singing of hymns, a reading of a chapter from the Bible, prayers, and a sermon, all of which Inglefield seemed to enjoy (pp. 24, 25). Members of the Esquimaux also came near the ship on several evenings and sang "native melodies that were … graceful and full of harmony (p. 26).'' Inglefield then allowed the crew to "go on shore for a walk" which provided them a change of pace and a chance to move about on land (p. 26). As they sailed away in the evening the Esquimaux women cheered the ship, which was reciprocated by "the sailors similar loud tokens of farewell and good feeling (p. 26)."

Several other collective events are also reported. Later in their journey, they make contact with an English ship and Inglefield and the doctor are invited to share a luxurious dinner with the captain and officers of the boat (p. 87). After that they are invited by the Danish Governor to a party and dance at his house held by the Esquimaux (p. 105). Several officers and crewmembers attended this event with the captain. And guns and rockets are fired from their ship for the purpose of entertaining everyone (p. 106). During the party, which all who attended greatly enjoyed, they cheered the King of Denmark, danced for several hours, and partook of a punch that was prepared for the event (pp. 105, 106). At the end of the gathering around midnight "some Esquimaux melodies were sung by the party" and the Governor sang the "Danish national hymn (p. 108)." Finally, the Esquimaux women "serenaded" the men as they returned to their ship (pp. 108, 109).

Two other occurrences also deserve comment. Both involve shared activities or reactions of the crew, in which their spirits are clearly boosted. After battling a horrible gale for about a week its end was collectively celebrated with the Sunday divine service and the captain allowing the men to go onshore to walk, explore the islands, collect items of interest, and visit the Esquimaux (p. 103). The day was "thoroughly enjoyed" by all (p. 103). The high morale of the crew was further evident, if not increased, when a little later in their mission the captain concludes that due to the impending winter, dangerous coastal conditions, and unrelenting bad weather it was time to end their journey and return to England. His decision is met with cheering, the apparent happiness and relief of the men, and the captain's thanks to the Almighty for the success of the mission (pp. 112–114).

Inglefield also refers to one additional RSP. A common practice involved the naming of points of land, islands, mountains, and bays. The names used were normally famous or noteworthy officials, royalty, friends and acquaintances, and political figures. The captain would usually choose the names (pp. 69, 70, 71, 76). However, at one point he uses "we" to discuss the naming of "certain headlands," which suggests that to some extent others, i.e., officers and/or crewmen, participated in this ritual practice (p. 80).

In total, while other rituals such as the captain reading in his cabin are periodically referred to, religious practices, work and scientific activities, shared group events and social interaction, and the naming of geographical sites are the four major types of RSPs regularly engaged in by the members of the mission (p. 94).

To conclude, all available evidence indicates that the crew's morale was generally high throughout their journey and there were no problems, schisms, or conflict among any of the men even though they encountered many difficult situations involving bad weather and threatening ice floes. During their voyage they consistently participated in a number of RSPs. These activities were meaningful to the men and certain ones were of great importance, i.e., high rank, such as religious practices that occurred throughout the mission. Their high rank was a product of their salience or prominence in the lives of the crew, their frequent occurrence, their similar or homologous nature, and having the resources to perform them, e.g., bibles and religious materials, areas on the ship where they could collectively meet to worship, and access to Esquimaux/Danish religious services on land. Moreover, they were able to engage in several very pleasurable collective events such as a party and dance marked by strong shared emotions.

It is important to keep in mind that their participation in RSPs, good relations, and positive state of mind was not due to simple chance. Quite the contrary, they were the result of several factors. First, the sound, effective leadership provided by the captain was crucial to the mission. He was dedicated to maintaining the high spirits and equanimity of the crew including his officers. And he promoted various dominant RSPs as has been discussed. Second, the short nature of the trip, four months, certainly worked in their favor. They did not have to endure the difficulties of living on their ship for a long period of time during one or more Arctic winters. Third, they had several opportunities to visit and share ritual practices with people at a settlement and on a different ship. In marked contrast, many expeditions were deprived of much, if any contact with other human beings. And fourthly, the crew was English, which meant that their common ancestry and cultural background created a shared familiarity with, if not commitment to, certain religious beliefs and experiences. This made it easier for crewmembers to engage in shared religious and other rituals during their journey. The result was a successful expedition in terms of the personal and social health of all on board.

We now turn to another Arctic venture, although quite different from the preceding expedition and for that matter all of the other missions examined in this study. The group in this land journey involves just two persons. However, the social and ritual dynamics are in crucial ways not that different from what we find in other successful endeavors.

A Land Journey in the North: The Team of Haig-Thomas and Nookap

David Haig-Thomas's (1939) memoir, *Tracks in the Snow*, describes a mission (1937–1938) that was funded by several British firms. While the key part of the endeavor involved a land journey that occurred approximately from the winter to the early summer of 1938 the two men faced great hazards and difficulties. Educated at Cambridge University, Haig-Thomas and his Eskimo companion, Nookap, sledged in Northern Greenland and the polar terrain beyond to an unexplored island, Amund Ringes, in search of prehistoric animal skeletons. While the two never found any fossilized bones, they still engaged in a remarkable undertaking. His account contains extensive pre and post discussion of the expedition, but we focus here on the sledging trip itself where the two explorers formed a tight-knit unit.

Land journeys such as this one involving dog sledging present travelers with an array of unique challenges. Naturally they are exposed to the elements including bitter cold, snow, winds, and storms. These conditions can be worsened, however, by other factors such as bad ice in which rough ice is buried in snow. It requires great exertion when a sledge gets stuck and a person sinks in the loose and deep snow, which then causes heavy perspiration and one's clothes becoming soaked. Even removing heavy clothes may not prevent perspiration. This can then lead to extreme cold and chills, both while sledging and even in the tents or igloos the two men built when they tried to rest (pp. 172, 173, 180, 182).

A different type of difficulty involves animals stealing their food supplies. This happened at least twice when the two left a cache of food buried in the snow and in an igloo for their return trip from the island where they went to search for fossils. To their dismay, bears had found and eaten all their food supplies, which made this part of their journey quite exhausting and even more perilous for both the men and the dogs that had to be fed (pp. 224, 225). And other complications can arise, sometimes unexpectedly, as when Haig-Thomas found himself covered in human lice, which was extremely irritating until he placed all his underclothes by one of his dogs that slept outside the igloo at night. This killed the lice while not disturbing the dog (pp. 226–228).

More generally, ever-present dangers on land journeys such as this range from simply running out of food to being attacked by animals such as bears or wolfs who do not know how dangerous humans can be when armed as Haig-Thomas and Nookap were. And there is always the possibility that, while sledging one may break through a weak spot in the ice or snow and fall into a deep hole, crevasse, or icy water all of which would probably be fatal.

Concerning the mental and emotional state of the two men and relations between them, we find almost no references to personal problems, low morale, conflict, the absence of RSPs, or their ineffectiveness. On one occasion they did have a slight disagreement about whether Haig-Thomas should keep a bottle of Schnapps or get rid of it as Nookap wanted (p. 177). And another time they had a minor difference of opinion about whether Haig-Thomas should have done a better job of controlling his dogs when they ran off after a hare. But that ended quickly when the author admitted to his companion that he lacked the skills of an Eskimo and the latter visibly displayed regret over his remarks. Haig-Thomas himself states: "This was the nearest we got to a quarrel on the whole of our trip (p. 185)." In reality, they two never experienced any serious troubles in their interaction with each other or suffered from any type of conflict during the entire journey.

Throughout his memoir the author does comment on a variety of RSPs engaged in collectively and individually by the two, along with a number of observations about their state of mind and interpersonal relations. One joint activity involved different types of conversations and discussions such as Nookap telling stories about other white men who had taken bottles of liquor on sledges and how they were broken, the two men describing their dreams and interpretations of them, their talking about and debating the quality of previous expeditions by others and discussing Eskimo religious beliefs and taboos (pp. 177, 181, 181, 182, 189). Other kinds of storytelling involved Nookap describing to Haig-Thomas an encounter he once had with a bear that he shot and killed along with other past experiences with bears (pp. 196, 197). On other occasions they discussed topics such as how white men and Eskimos remember something, Nookap's dream that the author had been killed by wolves, and several stories Nookap told about how different Eskimos had come close to being attacked by wolves. They also talked about the sugar and tea they had left behind at a camp. And on their return trip they met an Eskimo bear-hunting party who they greeted and exchanged stories with (pp. 231, 234, 243, 252).

Another kind of shared behavior involved the observance of appropriate ritual behaviors around a grave they found such as Haig-Thomas not examining or disturbing the items buried with a dead Eskimo hunter (p. 190). Finally, a very different type of collective experience occurred after the two had reached the unexplored island of Amund Ringnes and the author claimed it for the British Empire. Haig-Thomas states: "Nookap and I behaved almost as if we were drunk. We had another meal of bear and fired one shot from our 0'22 rifle. Just why I wasn't sure; but I felt it was the custom to fire some sort of salute or something (p. 220)." While not totally certain why they did this, they nevertheless felt compelled to share in this ritual celebration of their accomplishments.

Work rituals were also a significant part of their journey. Besides the almost daily dog sledging and periodic walks they took to explore their surroundings there were a number of other work-related activities they engaged in. For instance, when they camped one day Haig-Thomas says we "spent many hours filing and polishing our sledge-runners (pp. 187–189)." The procedure for doing this first involved using a course file to remove large dents in their steel runners followed by finely polishing them to make them extremely smooth. Later he and Nookap engaged in another task when they drew "a rough sketch-map of the coast of Amund Ringnes Island (p. 217)." Other jobs also had to be routinely carried out whether that involved "sewing our boots which had now got worn on the ice, mending harnesses, and putting new lashes on the whips" or hunting for game as when the author took a long walk and "shot three hares (p. 232)." Trying to find food by hunting was a common and very important practice.

In addition to these RSPs both individuals practiced a number of more personal rituals. Some of these were work related so the line between personal and work rituals fades in certain cases but for the sake of clarity we use the term "personal rituals" to identify these other activities. For instance, the author took photographs on a regular basis as is apparent in the pictures provided in his memoir and comments such as his wanting to take photographs of "musk-oxen" while hiking (p. 178). Later he refers to writing in his diary, which certainly was an often-repeated practice that provided the basis for his published memoir about their journey (p. 197). In contrast Nookap, who couldn't write, made his own contribution to the author's journal by drawing a picture of his encounter with a hungry bear (pp. 197, 198). Last of all, Haig-Thomas reports that one day toward the end of their travels while Nookap "spent a hard day getting up to the ice-cap," probably for hunting, course-plotting purposes, or some other unstated reason, he spent a lazy day with his "camera and glasses watching the musk-oxen (pp. 238, 239)." After hiking several miles, he found a herd made up of both adults and calves that he crept close to and observed.

Intertwined among these references to RSPs are ample amounts of evidence concerning the relations between the two men. Simply stated their ability to successfully work together almost every hour, day, and night, for the entire journey attests to their positive frame of mind and congenial relations. Their ability to carry out so many shared activities ranging from conversations and storytelling to work rituals shows they communicated and cooperated with each other, i.e., they formed a cohesive social unit and enjoyed each other's company.

These qualities are also evident in various comments by the author about their mutual situation. As already mentioned, he clearly states that while they had a disagreement one time about his handling of his dog team that was the closest they ever came to an argument (p. 185). His sensitivity to the feelings of Nookap are also apparent when he did not exam the remains of an Eskimo grave because his companion gave him a "disapproving glance (p. 190)." And his previously described celebration of the discovery of a new island where they behaved as if they were "drunk" was obviously marked by shared feelings of happiness and triumph (p. 220). Later in their mission, Haig-Thomas's admiration for his companion is also quite obvious when he refers to: "The great Nookap, who forgets nothing, and is probably the greatest Arctic traveller of all times … (p. 237)." Finally, toward the end of their journey the author provides the following assessment about the expedition as a whole:

> It had been a failure, for I had been unable to discover the prehistoric skeletons which were really the sole reason for the trip; but I found I didn't mind in the least. I had made a long sledge journey and enjoyed myself tremendously watching the animals. I had innumerable notes … I had also found a barren island of some two hundred square miles … (pp. 250, 251)

While not specifically mentioning his companion, one must think that his satisfaction with the mission was also the result of his successful partnership with Nookap, which made possible these very accomplishments.

In closing, one must ask what factors contributed to the social psychological success of this mission. While probably not exhaustive several points surely come to mind. To begin with, it was a relatively short journey lasting several months, which meant they did not have to endure for a more prolonged period of time the dangers and disruptions that accompanied such a venture. At the same time both men were experienced in Arctic travel and exploration. They knew first-hand the nature of the challenges they faced and the importance of working together if they were to succeed in this hostile environment. Moreover, the evidence indicates that they were open-minded in that they respected and liked each other. Along with this they possessed a strong motivation to accomplish the task they had chosen and were hard workers. Haig-Thomas was also a very effective leader, or perhaps we should say titular leader given the critical skills and contributions of his partner. The author at the very least was quite talented, made sound decisions, and understood the value of working with and participating in various RPSs with his partner (and by himself), much of which was grounded in his appreciation for the exceptional abilities of Nookap.

Finally, this expedition also shows how RSPs can be seemingly simple in nature, e.g., conversations, storytelling, repairing travel equipment, or observing animals, while yet being quite valuable in extreme situations. This is especially the case when they involve highly ranked RSPs that are salient and repeated often as was the case for these two persons. Many of these ritualized behaviors are also homologous, because they often emphasize the idea of interacting and working closely together in a cooperative manner. And the two travelers possessed various non-human and human resources ranging from tools and materials such as sledging equipment, a camera, and a diary to socio-emotional skills that support these RSPs and strong interpersonal relations. Simple rituals whether they are social or more personal in nature can be extremely consequential when they are meaningful and significant to those who practice them.

A Quest for the South Pole: The Tragedy and Triumph of the Scott Expedition

Our next case involves a very famous mission, the *Terra Nova* expedition to Antarctica spearheaded by Robert Falcon Scott, that while marked by great obstacles and tragedy was still a positive experience in terms of the personal and social equanimity of the crew. Scott did most of the fundraising to support this venture. Part of the funding came from various private businesses and backers some of whom were members of the crew. Half of the funding also came from a government grant while additional revenues were secured from loans. The crew of 65 men were mostly British (many current or former British military) with a few exceptions such as a Norwegian ski expert.

Our sources for investigating the expedition include the memoir *The Worst Journey in the* World written by the Englishman Apsley Cherry-Garrard ([1922] 1965) who served as an assistant zoologist on the mission. While opinions vary some consider his book to be one of the finest accounts of a real-life travel adventure. The expedition and author have also been depicted in film and television programs. The second source is Scott's own diary *Robert Falcon Scott Journals: Captain Scott's Last Expedition* (2006). Together they provide a great deal of information about an expedition that began in 1910 and ended in 1913 whose goal was to be the first persons to reach the South Pole. Ultimately, Roald Amundsen, the famous Norwegian explorer, reached there first.

Cherry-Garrard's account has nearly 600 pages and draws upon his and other accounts, usually from crewmembers' diaries and notes. Scott's diary has approximately 500 pages. Given the large amount of information contained in these sources it is impossible to conduct a detailed documentation of every occurrence on the mission. Nonetheless, because we draw upon two major sources our discussion will be longer than others in this chapter.

Given the length of the expedition and the challenges the men faced, the mission had several stages some of which occurred simultaneously involving different members of the crew. Actually, the expedition was composed of at least half a dozen parts. They included a five months long voyage by ship from England to Antarctica, setting up the camp after their arrival and establishing depots with food and fuel at various points on the projected course for the group that would eventually try to reach the South Pole, an arduous journey by Cherry-Garrard and two others – Wilson and Bowers – to Cape Crozier dragging their sledges by themselves over rough surfaces and bitter cold to find Emperor penguin eggs (this phase of the expedition was the source for the book title *The Worst Journey in the World*), and Cherry-Garrard being in one of the support groups that went part of the way with the main team made up of Scott and three others who would ultimately make the attempt to reach the South Pole.

Other major phases of the expedition included Cherry-Garrard and another crewmember who travelled to "One Ton Depot" with extra food to hopefully meet the Scott team on their return from the South Pole (after about an unsuccessful week's wait the two men returned to the base camp), the last part of the mission involving scientific and other work at the main camp and a search party the author was a member of to find the Scott team, and lastly, the experiences of Scott and his companions on their journey to the Pole and their attempted return.

We should point out that while Cherry-Garrard was at times dejected and endured certain physical problems the crew's spirits remained relatively high a good part of the time. This was the case despite many ups-and-downs with storms, equipment troubles, and the tragic loss of several persons.

More precisely not only was the crew isolated from the rest of the world for an extended period of time; they experienced many other problems. For instance, Cherry-Garrard reports that during their ocean voyage they encountered a gale, had to toss sacks of coal overboard, and saw their ship begin to fill with water before it was pumped out (pp. 49–58). Upon reaching Antarctica in January 1911, while establishing their main camp and setting up depots on the path to be taken by the South Pole team, they found themselves working through exhaustion with little time to sleep and eat, beset by a three-day blizzard, and having to deal with part of the crew becoming stranded on broken floating ice blocks (pp. 104, 113, 140).

Then during the trek to Cape Crozier in search of penguin eggs the author and his two companions experienced such extreme cold and blizzards they were deprived of sleep which took a mental toll on them (pp. 241, 274, 275, 279). In fact, Cherry-Garrard became so cold that the nerves in his teeth died and all his teeth were "split to pieces (p. 292)." Such developments led the author to describe their journey as a "horror (p. 298)." In the subsequent period when preparations were being made for the final push to the South Pole, a number of disruptions also occurred including a crewmember freezing his hand, a man falling from an iceberg and suffering a concussion, a knee injury to one of the crew while playing soccer, a dog dying due to unknown causes, and a mechanical accident involving a piece of equipment (pp. 309, 310).

Notwithstanding the author's discomfort and sometimes dissatisfaction with conditions we still find throughout his memoir numerous comments about a variety of ritualized behaviors engaged in by crewmembers, along with a large number of remarks about their good spirits and interpersonal relations. These comments concern both Cherry-Garrard's and other's experiences. Many times, his references to RSPs are interlaced with descriptions of the crew's positive state of mind, their dedication, and compatibility.

On their voyage to Antarctica, for instance, there was a ritual ceremony celebrating crossing the equator marked by merriment, roughhousing, and singing, a practice they often engaged in (p. 10). Comments are also made about celebrating with a champagne dinner after they finished moving all the coal in the ship's boiler-room and singing songs while removing water from their ship during a gale (pp. 34, 51–56). Scott would sometimes also hold a religious service on Sundays (p. 38). More often the men engaged in work rituals, sometimes involving very strenuous activity. Reflecting the high morale on the ship "sandwiched" into this work was a "considerable amount of play" and "joking (pp. 44, 82)." Cherry-Garrard also says that when Christmas arrived, he did not think, "many at home had a more pleasant Christmas Day than we (p. 76)." The big day began with church and hymns, followed by decorating the work room with sledging flags, a Christmas dinner of fresh mutton and penguin, toasting, much singing, and banjo playing (p. 76). So too, New Years was celebrated by a happy "rowdy mob (pp. 80, 81)."

At the center of all of this activity and the perilous times when bad gales threatened the ship was their leader Scott who is referred to in extremely approving terms as "simply splendid" and "one of the best (pp. 54, 56)." The author's assessment of their leader and the crew as a whole is supported by another highly accomplished, respected crewman, Bowers, who states this "is a good place to live in" and describes the crew as a "cheerful and ribald lot of Souls (p. 56)."

In addition to the author's extensive diary writing many other comments are made about ritual behaviors and the spirits and camaraderie of the men (p. 87). For instance, much of their time involves work practices such as collecting water samples, taking photographs, and other scientific tasks in addition to the basic chore of getting the camp up and running (p. 131). Cherry-Garrard states: "The ordinary routine of scientific and meteorological observations usual with all Scott's [missions]" was followed (p. 181).

So too, other sorts of RSPs are engaged in. One crewman states that he would often go for walks with Captain Scott and discuss different topics such as his plans for the mission while Cherry-Garrard refers to the pleasure of skiing (pp. 154, 192). We also earn that they would spend their days when weather permitted hunting and butchering seals which were used for fuel, light, and food, climbing nearby craters and hills, and engaging in long discussions in the evenings (p. 161). Moreover, the author describes how the men would routinely carry out discussions around the blubber (seal fat) stove at night that were both intellectual and comedic and most were glad to listen to and participate in the conversations, which involved everyone including the leaders (p. 163). He also tells us that reference books and other reading material were in great demand (pp. 163, 164). And some individuals pursued more personalized activities such as a crewman named Wilson who often sketched and made watercolors of the natural scenery.

Both collective and individual RSPs were quite popular. For instance, a small team of men who were carrying out a task away from the main camp spent a happy Easter Sunday in a howling blizzard, enjoying a special dinner of canned haddock, biscuits, cheese, pemmican, and cocoa with milk followed by reading (p. 175). At the same time, the author emphasizes how important "plenty of out-door exercise" was, accounting for the "happiest and healthiest members of our party (p. 193)." This morale boosting practice was often done alone he says, not due to a dislike of others, but because it provided a welcome change from living closely with everyone most of the time. In a different vein, he also notes how uplifting to their spirits it was to have a pianola which was routinely played by different persons before their collective suppers (p. 197). After supper pipes were often smoked and sometimes gambling occurred where cigarettes or a dinner when they returned home were used as currency for their bets (p. 196). And the author talks at length about how in the evening people enjoyed playing games such as chess or backgammon, doing various jobs, listening to their gramophone and records of classical music, and reading books, their collection of which included modern fiction and Antarctic and Arctic travel (pp. 198, 199).

Moreover, Cherry-Garrard describes a very popular lecture series that was held in the evenings. The first winter the lectures were held three times a week and the second winter twice a week (pp. 212, 213). The men, including Scott, lectured on a wide array of subjects ranging from sledge foods, polar clothing, one individual's journeys to Tibet, ice barriers and inland ice, meteorology, geology, and sketching to the mismanagement of horses (pp. 212, 213, 214, 215, 217).

Interwoven among these references to RSPs are numerous comments about the spirits of the crew and their relations with each other. While some statements are made about poor morale such as people being depressed when finding an old hut in a desolate condition, becoming disheartened, "cold and depressed" by a blizzard, and later feeling "dark, gloomy" and depressed during a difficult march the large majority of commentary emphasizes the rituals people engaged in, their upbeat feelings, and harmony between the men (pp. 97, 98, 136, 137, 139).

For instance, we learn that shortly after landing everyone works together in a "splendid way," "everyone is happy," and "Scott seems very cheery about things (pp. 88, 94, 96)." As they become more settled an abundance of statements of the same nature follow including references to an "amusing argument," being "quite happy" after erecting a camp on a difficult depot trip, making a "cheering drink" from melted snow, and having "great fun" sledging downhill (pp. 118, 121, 123, 151). In discussing their hard work of making a "path to the Pole" that the smaller team will eventually take, the author also makes the striking observation that: "Our sense of co-operation and solidarity had been wrought up to an extraordinary pitch; and we had so completely forgotten the spirit of competition [to reach the South Pole first] that its sudden intrusion jarred frightfully (p. 129)." In addition to the aforementioned happy Easter dinner during a raging blizzard and enjoyable evening discussions around the stove, we also learn that incidents such as cooking failures were recalled with amusement and remembered as "outstanding features of our existence (p. 164)." And we are told that ever-present work routines created a sense of normalcy among the men (p. 181).

And so it goes whether commentary involves enjoying the "food, bath and comfort" of their camp, having a "pleasant" lunch and benefiting from an "atmosphere of pleasant and quite interesting conversation," or a lengthy discussion about the importance of listening to music on a gramophone, playing games such as chess and backgammon, and reading modern fiction and books on Arctic/Antarctic travel and related information (pp. 181, 195, 198–200).

Cherry-Garrard also provides a lengthy description of the officers especially Scott, Wilson, and Bowers (pp. 200–212). Concerning their leader, he emphasizes that Scott was the "driving force of the expedition," "the most dominating character" in the group, and had a great strength of "character" and "sense of justice (pp. 200–202)." And even during the author's extremely difficult and painful trip to Cape Crozier with Wilson and Bowers he reports that they were "cheerful," sang songs, and never complained nor failed to stand the test of "self-sacrifice (pp. 246, 296)."

Finally, in reflecting upon the entire three-year mission, he states there was never "any friction of any kind" among the crew (p. 225). The reason he says was due to their strong motivation to work and obtain scientific results (pp. 225–229). As such they "had no idle hours: there was no time to quarrel (p. 226)." In other words, much of their time was consumed by RSPs involving work centered to a very large degree on their scientific objectives.

Nevertheless, the overall tone of the undertaking does change in the latter part of the expedition when the men prepare for the trip to the Pole, different teams accompany Scott and his three companions part of the way on their journey, the crew wait for the four-man team to return from the Pole, and they search for the team. For instance, statements are periodically made about how "the camp is silent and cheerless," "whole days passed without conversation," how "depressed" all their diaries became when they had bad weather, and how it was "sad saying good-bye" to the four men – Scott, Bowers, Wilson, and Oates – when they began the final push to the South Pole (pp. 324, 331, 337, 381). At the same time, we also learn that the "routine of a tent makes a lot of difference," the routine is a "comfort," they keep their "spirits up," and while sad they still shook hands, made their emotional good-byes, and cheered the Polar team when they departed for the Pole (pp. 329, 330, 354, 385).

It is not that surprising to see a change in the mood of the crew given that they are now in the latter part of a very long mission, and they have finally reached the very dangerous stage where the four-man team is on its own. Yet, the rest of the group in the camp continued to work and perform their duties, engaged in a number of RSPs, did not fall prey to conflict, and performed several meaningful rituals when they eventually find Scott and the others.

During their last winter, Cherry-Garrard says that the crew decided, "as far as possible, everything should go on as usual" which meant that scientific work was continued, their dogs and mules were cared for, and so on (p. 437). They also decided to write a second volume of the "South Polar Times," a self-produced magazine with various articles and observations by crewmembers including an editorial by the author (pp. 437, 445). All of this was done to prevent "any sense of depression" becoming "part of the atmosphere of our life (pp. 437, 438)." Efforts were also made to exercise and take walks when the "constant blizzards" allowed them to do so along with other activities such as competing with each other by fishing through holes in the ice (pp. 438, 444). And then there was the lecture series, which continued at a pace of twice a week.

Generally, all these activities were less enjoyable due to this being their second winter trapped inside much of the time due to the weather and the fate of Scott's group unknown, although by this time they were in truth, certain of the demise of the polar team. Still, the author notes that one of the lectures dealing with snow surfaces and their effects on sledging was "especially interesting (p. 455)." In addition to their determination to remain active by engaging in these RSPs, evidence also indicates that the men were able to communicate and cooperate with each other as when the whole party met and nearly unanimously decided upon a plan and direction to search for the missing team (p. 443).

Their search led to the discovery of the missing group who reached the South Pole and died on the return trip about ten miles from the base camp (p. 480). The frozen bodies of Scott and two others, Wilson and Bowers, were found in their tent while a fourth individual, Oates, who was sick had walked to die in solitude in the hope that the others could make it because he was slowing their progress. The importance of ritual is quite evident in the way the search party handled the discovery of the three deceased men. A burial service was held which included prayers and a reading from a chapter in Corinthians (p. 481). Following written instructions left by Scott his diary was read along with his "Message to the Public" and account of Oates' death (p. 483). A "Cross [made of skis] and Cairn" were then erected over the bodies accompanied by prayers (pp. 483, 484). A signed note was also left with the memorial cairn and cross that included a tribute to Oates who had walked away to die (p. 484). While profoundly saddened by the loss of their comrades, these burial rituals and the other RSPs they engaged in undoubtedly helped the men cope with their situation and continue to function as an effective group.

Finally, the author provides some very useful information about the experiences of the four-lost explorers during their trip, which was based on the diaries of the men, especially Scott's journal. Not surprisingly, as conditions and their physical health declined during the return journey the men's morale suffered as reflected by the author's observation that "All the joy had gone from their sledging (p. 515)." In reality, a variety of factors contributed to their ill-fated return including insufficient fuel, extreme cold, caring for an ill crewman, lack of food, and unexpected, exceptionally harsh sledging conditions. Still, as the situation became increasingly grim and hope must have faded, the men exhibited what by all accounts would be considered admirable behaviors remaining active and participating in different ritual practices until the very end.

We find ample commentary concerning what they did, said to each other, and how they felt which supports this assessment. Various remarks reveal that they continued to engage in work, exploratory, and scientific rituals such as taking photographs in addition to collecting rocks, making geological observations, and continuing their sketch work (pp. 507, 521, 522). Moreover, Scott (2006) tells us that they shared several enjoyable dinners on Christmas day before their food supplies ran out involving "luxuries, such as chocolate and raisins at lunch" and a "feast" at dinner, which included pemmican and biscuits (pp. 359, 360). Later, he describes another comfortable dinner in January as "a fat Polar hoosh" made of stew, chocolate, and a cigarette (p. 376). They were able to enjoy the dinner even though they knew that they were not the first to reach the South Pole since Amundsen had got there before them. Other references are also made in March to their positive morale, or at least their attempts to maintain the appearance of good spirits, when Scott reports that while the situation is quite bad, they "still talk of what we will do together at home," they "constantly talk of fetching through," and "The others are still confident of getting through – or pretend to be – I don't know (pp. 407, 411)." Cherry-Garrard ([1922] 1965) also confirms that Scott says several times they are continually cheerful and talk of making it (pp. 532, 533, 537, 538).

Even in the last days and hours before he dies Scott's (2006) behavior is telling for he continues the RSP of writing in his journal detailing the final experiences of the team in addition to writing letters to various family members of the crew and others he knows where he refers to the good qualities and spirit of the group along with a Message to the Public (pp. 415–422; see also Cherry-Garrard [1922] 1965, pp. 538–542). This is a very meaningful practice to Scott (2006) as seen in his next to last sentence when he says, "It seems a pity, but I do not think I can write more (p. 412)."

Still, most of Scott's diary deals with the period prior to his South Pole trek. In this respect while Cherry-Garrard's account is well known and covers many aspects of the expedition, Scott's diary is very, if not in certain ways more informative about the daily RSPs and dynamics of the group prior to his departure for the Pole. Numerous entries discuss the everyday activities performed by crewmembers.

To avoid too much overlap with our previous discussion of Cherry-Garrard's account we simply identify the separate RSPs Scott describes often multiple times. They are: work and scientific rituals, divine services and hymns, playing or listening to music, special dinners and celebrations, the preparation and reading of several issues of the Polar newspaper, pausing to have tea, walks and skiing, playing football, exercising, conversations and smoking, writing letters, playing games such as chess, and lectures. While all are important, some standout such as repeated references to the work rituals of the crew, the lecture series, and religious services (for a description of a typical Sunday including services see pp. 212, 213). Concerning the presentations, we find in Scott's diary 31 comments about this activity, a very high number compared to other expeditions examined in this study many of which did not even hold lectures.

Scott is quite perceptive about the value of regularly engaging in practices such as these in terms of their impact on the social-psychological dynamics of the crew. Relevant commentary is often interlaced with discussions of the types and quality of ritual activities performed by the crew, the effect of these practices on the social equanimity and state of mind of the group, and his observations about the importance of these behaviors and making the right decisions as a leader. For example, at one point he states, "The busy routine of our party" is "very good" and "a subject for self-congratulation (p. 198)." He also stresses that such routines depend on "forethought" and "preparation" which should result in the proper materials being brought on an expedition, i.e., what SRT refers to as resources that are necessary for the successful enactment of RSPs. Later, Scott provides a detailed description of the highly valued "daily routine" he has organized, which creates a "settled regularity" for the men (pp. 227, 228). It includes times for rising, going to bed, meals, scientific work, a host of tasks and odd jobs, and evening activities ranging from work to reading, playing games, writing, and evening lectures all of which result in a life that is not "idle (p. 228)."

So too, Scott discusses a number of times the high morale of everyone, their camaraderie, the lack of rifts and conflict, the absence of signs of depression, their commitment to their work, and the lack of problems among crewmembers (pp. 44, 60, 176, 193, 207, 216–218, 232, 233, 253, 285, 286, 303, 304). The amiable and cohesive nature of the crew even leads him to state that "The study of individual character is a pleasant pastime in such a mixed community of thoroughly nice people, and the study of relationships and interaction is fascinating … (p. 303)." He then goes on to say, "I think that it would have been difficult to better the organization of the party - every man has his work and is especially adapted for it … it is all that I desired, and the same might be said of the men selected to do the work (p. 304)."

All told from the perspective of SRT, ample evidence shows that the crewmen regularly engaged in RSPs that were salient, highly repeated, complemented each other in focusing on and/or supporting the work and amicability of the crew, and were usually facilitated by ample resources. These activities also included some collective events, such as birthday parties and special dinners that were infused with warm feelings and a spirit of congeniality. In many ways, these practices were the result of the planning and decisions of a very effective leader, supported by other key individuals. What resulted was a group of individuals who were united, cooperative, communicated well with each other, and shared positive feelings with their fellows. They also possessed much of the time a good outlook about their state of affairs.

At the same time, there were ups and downs throughout the mission that impacted how they felt about their situation. This was especially the case in the last year when Scott's team took their final journey, failed to return, and was finally found, a period when the crew's morale was lower. And during the Crozier trip, the three men faced punishing conditions, which Cherry-Garrard suffered from and held a very negative opinion of (as reflected in the title of his memoir). In total, this was a successful venture, which was sometimes marked by high degrees of success in a social psychological sense, great efforts, and truly commendable behavior.

First to the North Pole: Henson's Account
of the Peary Expedition

Our next venture, led by Commander Robert E. Peary, is another famous expedition, which reached the North Pole. In addition to Peary, the team was composed of eight men from the United States, Canada, and Greenland. The mission's sponsor was the National Geographic Society, which funded many of Peary's expeditions. While debate and disagreements have continued for over a century about who actually reached the Pole first – Robert E. Peary or Frederick Cook – the general consensus is that Peary did, although we may never know for certain.

Our source for examining the Peary expedition (1908–1909) comes from the first African American polar explorer, Matthew A. Henson, author of *A Black Explorer at the North Pole* ([1912] 1989). Henson was a very experienced explorer who accompanied Peary on eight polar missions between 1891 and 1909. He was a highly accomplished Arctic traveler who possessed essential skills such as knowing how to drive dog teams, build igloos, and repair sledges, in part because he had developed over many years a very close relationship with a number of North Greenland Eskimos whose language he had learned. They, in turn, held a very high opinion of, if not revered, Henson. What is more, Peary acknowledged that "I can't get along without Henson," who has been credited by others as being the most valuable man in Peary's quest for the North Pole (pp. xvii, xviii).

The journey was essentially composed of three stages: the ocean voyage from the United States to Northern Greenland, which receives less attention in Henson's memoir, the group's activities once they reached their destination, including preparations for the sledging trip to the Pole, and the final 400-mile journey which involved several support parties who accompanied Peary and five others who completed the final leg of the journey. We should appreciate that of nine men, six knew each other quite well, since they had worked together on previous Arctic missions with Peary. The remaining three also turned out to be amiable, hard-workers who got along with the others. And some of the Eskimos who assisted the group had travelled on previous expeditions with Peary and Henson. Actually, the final leg of the trek to the North Pole involved six men, Peary, Henson, and four Eskimos, several of whom knew both Henson and Peary, although formal recognition has by and large only been given to Peary.

Their quest for the Pole involved numerous problems. Miserable conditions were common throughout the venture including their ship, the *Roosevelt*, encountering a "blinding storm of wind and snow" and "impassable ice," some of their dogs dying from a mysterious disease once they reached land, and "frequent storms and intense cold" whose effect was "abject physical terror (pp. 32, 44, 47)." For instance, at one point Henson describes how he had to solder some tins which held their alcohol fuel while he was simultaneously "freezing to death" and nearly being "burned alive" due to the darkness, extreme cold, and poor "furnace arrangements (p. 54)." Other commentary also emphasizes how everyone including the "men of iron," the Eskimos, suffered from the intense cold and gales (pp. 59, 70).

Once they began the 400 mile trip to the Pole, most of which occurred on the frozen Arctic ocean, more difficulties befell the men including having to repair broken sledges in bad weather, Henson sometimes not being able to sleep or write in his diary due to the extreme cold, people becoming frustrated due to delays in the journey, and individuals being nearly killed by disintegrating ice floes when they are sleeping (pp. 78, 79, 80, 81, 88–91, 121). Furthermore, MacMillan, a crewman and professor from Massachusetts, had to return to the ship due to health problems (p. 95). A little later in the trek, Henson had to kill one of their dogs out of mercy because he fell into freezing water and became dreadfully ill (p. 117). Not long after that he himself fell off the ice into the water and nearly died save for his Eskimo partner who pulled him out (p. 131). Even worse, when one of the support parties had to return to the ship, a crewman, Professor Martin, accidentally fell into the cold Arctic Sea and drowned (pp. 114, 115, 148). After this the six-man group led by Peary completed the final leg to the Pole only then to face the difficult return trip to their waiting ship, a journey Henson describes as "a horrid nightmare (p. 141)." The nearly 17 days journey in the extreme cold, over drifting ice and ice ridges, was one of "haste, toil, and misery" and incredible "fatigue (pp. 140, 144)."

To say the least, these conditions presented a challenge to everyone's comfort, safety, and spirits. For instance, at one point the Eskimos' complaints about the cold became pronounced with two of the older men becoming so irritable, Peary sent them back to the ship and dismissed them from the entire expedition (pp. 90, 91). Still, notwithstanding these difficulties and rankled nerves, the group's overall morale remained relatively high. A large amount of commentary refers to the good spirits of the crew and positive social dynamics. And these comments are often interspersed with discussions of different RSPs engaged in by both individuals and the group.

The rituals practiced by crewmembers are fairly varied in nature with some exhibiting a high degree of intensity. Certain activities, while mentioned less often than others, were still of some significance for these travelers. For example, Henson reports that when they were unloading the ship after their voyage there was much singing, which made him feel good and how the rule that everyone should take a bath on a regular basis became "contagious (pp. 35, 38)." He also mentions several times the value of conversations and the company of others including sharing an igloo one night with MacMillan, laughing and talking with Dr. Goodsell, the physician, and on another occasion talking with Goodsell and joking with MacMillan (pp. 58, 63, 89). So too, he mentions several times doing a dance around their campfire to keep warm and another individual exercising (pp. 65, 71). And on a few occasions, he refers to religious practices i.e., praying that the weather would remain good while on their trek to the Pole and reading from his Bible, which he had neglected for some time (pp. 109, 114).

Other RSPs, however, occurred more often, were more prominent, and had a greater impact on the men. Henson wrote in his diary on a fairly regular basis (e.g., p. 54). And he mentions one of his companions, Goodsell, writing in his diary (p. 89). Reading was also important to some as emphasized by Henson when he states that on the ship there was an "extensive library" on polar topics in Peary's cabin and the author himself had a number of books that he spent "many a pleasant hour with" during the winter (p. 39). He also kept track of dates with a calendar he brought along (p. 39). And he says that on a sledging trip prior to the trek to the Pole some men took a few books along like a Bible and a small set of Shakespeare, and on the ship each member had some "favorite books in his cabin, and they helped to form a circulating library (pp. 65, 66)." On the long trek to the Pole, he also remarks that during a delay Goodsell "reads from his little books (p. 89)."

An even more consequential and ever-present RSP for everyone centered on their daily work rituals. Henson refers several times to these practices stating that "I have a steady job carpentering, also interpreting, barbering, tailoring, dog-training ..." and "I have been busy making sledges (pp. 37, 38, 40)." He then goes on to stress that during the winter while living on the ship "constant activity and travel were insisted on" by Peary as opposed to hibernating (p. 46). The crew was constantly busy preoccupied with activities such as going on hunting parties, Professor Marvin carrying out scientific work, sledging equipment and provisions to a site for the North Pole trek, and on-going work at their headquarters such as building sledges (pp. 46–47). Later, during the big trek, Henson shares a similar observation about their spending "the day industriously in camp, mending footgear, harness, clothing, and looking after the dogs ... (p. 122)." In essence work and scientific rituals such as collecting "natural history and ethnological samples" were a pervasive part of expedition life (p. xiv).

In a different vein, MacMillan often lightened the mood of the men and encouraged a number of recreational RSPs during their trip to the Pole. For instance, during their trek, he got everyone, including the Eskimos excited about a cache of equipment he had which presumably belonged to different persons, but it actually turned out to be a joke (pp. 87, 88). Henson tells us MacMillan was very good-natured and routinely joked with everyone. He is "the 'life of the funeral' and by his cheerfulness has kept our spirits from sinking to a dead level (p. 89)." He also created a very popular "series of competitions in athletic endeavors" such as wrestling, boxing, and stunts, which kept the crew, especially the Eskimos, active and preoccupied (pp. 89, 90).

Not long after this, the supporting parties gradually began their return to the ship. The overall high morale of the crew during this stage of the trek is quite apparent, as when Henson refers to good-hearted joking between a number of the men, Eskimo and non-Eskimo alike, and their heart-felt good-byes when Dr. Goodsell congratulates Henson and is applauded by everyone before he leaves (pp. 96, 108). When Professor Marvin had to return to the ship, he also congratulates Henson and then they "gave each other the strong, fraternal grip of our honored fraternity" [of Artic explorers] (p. 114). And when the ship's Captain Bartlett and his Eskimo comrades prepare to return to their vessel, cheerful goodbyes are shared, and Henson describes him as an "unconquerable hero (pp. 125, 126)."

As the six-man party continues, and finally reaches the Pole, a new RSP emerges, celebrating their accomplishments. They begin to give their camps names instead of simple numbers, e.g., Camp Nansen, Camp Morris K. Jessup (p. 132). Peary also begins to hang an old silk flag at the camps, a flag he had "carried on all of his Arctic journeys" and had flown at his last camps (pp. 132, 133). When the flag is raised, Henson says, "A thrill of patriotism ran through me" as he and the other men cheer (p. 133). He also states that the flag (and the symbolic and emotionally charged ritual focused on it) is "as glorious and as inspiring banner as any battle-scarred, blood-stained standard of the world;" it is a "badge of honor and courage (p. 133)." As on previous expeditions, a piece of the flag and a note are placed in a cairn (a tin) and buried in the ice by Peary in recognition of their achievements (pp. 133, 134). Peary also flies additional flags representing different organizations he is associated with, such as the Navy League and several college fraternities (p. 134). Shortly after that, he flies another flag at the "official" North Pole near the final camp (p. 136). As part of the celebratory ritual photographs are taken of the six men. Henson ends by recounting how moved he is by their achievement, and the ritual events surrounding it, when he describes the pride he feels being a member of the first crew to reach the Pole and being the first African American to be a part of this "great *work* (p. 136)."

Henson also comments on their high morale, cooperative spirit, and commit-ment to the venture, the group and their leader, and their very difficult return trip. Despite frequent grumbling and the two Eskimos leaving the expedition, he describes his four Eskimo companions, two of who had accompanied Henson and Peary on previous missions, in extremely respectful terms. He stresses how "devoted" and "blindly confident they are in the leader" and how they "worked faithfully and willingly" inspired by him (p. 137). He also says that Ootah, his closest Eskimo companion, and he had "saved each other's lives more than once (p. 138)." He then concludes by saying they exhibited "unswerving loyalty and fidelity" and had "remained steadfast and true (p. 139)." Attesting to the strong, effective leadership of Peary, Henson states that "we could not have gotten back without him" and "we never forgot that he was still the heart and head of the party (p. 140)." He had the author concludes, "taken the North Pole by con-quest" through the "tremendous fighting power of himself (pp. 141, 142)."

In sum, this expedition and the others in this chapter are strikingly different from failed expeditions in terms of the rituals engaged in, morale, and social conditions such as group cohesion and leadership. While facing great difficulties and the loss of one person's life, Henson's account indicates that a variety of RSPs were practiced by crewmembers, some to a more limited degree and others much more often such as work rituals, joking and recreational activities, flag flying and celebrating, diary writing, and reading. They were the most repeated, salient, resource supported, and similar rituals in that a number of them focused on the group, the basic needs of crewmembers, and the actions needed to ensure the success of the endeavor. At the same time, differences between this and other expeditions are also evident because celebratory events such as flag-flying activities involving shared emotions and collective pride occurred at the end. But no evidence exists that collective celebrations occurred during the expedition itself, e.g., Henson did not talk to anyone about his birthday and there were no festivities for the reappearance of the sun at the end of winter (pp. 28, 29, 86).

Finally, various factors facilitated the performance of RSPs, morale, and the unity of the group. Quite importantly, most of the men, like Henson, were experienced polar explorers and had previously worked before with Peary, thus, they knew what he expected and were committed to him and the mission. Henson also had excellent relations with the Eskimos, who provided essential support (and more) for the crew. Moreover, Peary was a very skilled and respected leader, so the men were motivated to perform their tasks, which contributed to the stability of the group and their eagerness to celebrate their achievement once they reached the Pole. MacMillan also had a major impact on the crew with his promotion of recreational rituals and his repeated joking with everyone. Finally, since most of the team members were experienced explorers, they were more adept at finding meaningful ways of spending their time whether that involved rituals of reading, diary-writing, conversing, bathing, or working.

In closing, it is worth mentioning that on the first encounter between Peary and Henson after their expedition, Peary never mentions the North Pole or their experiences. And the long-time, close relationship between the two ended after their 1908–1909 mission (pp. 153, xviii). Given their shared history such behavior seems puzzling but attempting to understand why, if that were even possible, would take us too far afield from our present concerns.

Research in Antarctica: A Mid-Twentieth Century British Expedition

We now turn to a much more recent expedition carried out between 1945 and 1948 that was part of the British Antarctic Survey focused on geographical and scientific study. While not famous the mission was conducted by a British male crew, who seemed quite cognizant of the effects of morale-boosting activities. *Two Years in the Antarctic,* written by E. W. Kevin Walton (1955), a World War II veteran and officer, describes life in their base camp and their various dog sledges which could take months to complete whose main goals involved surveying the environment, creating supply depots for subsequent travelers, and discovering suitable routes. Surgeon-Commander E. W. Bingham, whose skill and leadership the author speaks quite highly of, headed most of the mission during Walton's tenure.

This undertaking benefited in different ways from the experiences of polar explorers that came before them, along with improvements in the equipment used in the mission. We also find many ritualized behaviors occurring in this expedition, whether they involve familiar RSPs or new types of practices such as watching movies and using radios for communication.

While spared the tragic consequences seen in some expeditions, this mission still faced its share of disruptive events and problems. These challenges ranged from the difficulties of being separated from one's family to blizzards and heavy winds that sometimes kept crewmembers stuck for days or weeks inside their base camp or tents when travelling (pp. 19, 84, 88, 89, 92). Accidents and injuries also presented serious threats to explorers such as, when on a sledging trip a crewman fell into a narrow crevasse in a glacier (pp. 82, 83). Walton was lowered into the fissure and was able to rescue the individual although it took six months for damaged nerves in his hands and wrists to heal and the man to regain total control.

Notwithstanding these difficulties and the usual irritations that arise between people living together in such close quarters for long periods of time, we find many references to RSPs shared by crewmembers. To begin with, Walton describes early in his memoir how a "definite routine" in the camp promoted by their leader preserved "a smoothness of life (pp. 33–38)." He carefully details how the cook and others would rise at a certain time in the morning, eat breakfast, carry out a number of jobs including cleaning their home and scientific tasks, care for their dogs and kill several seals for dog food, eat lunch, return to work, feed the dogs, have their tea-time at 4:40 p.m., supper at 8:30 p.m., engage in recreational or work related activities, bathe, and then to bed around 10:00 p.m., to read and sleep. And he points out that their exceptional scientific library was of great help for carrying out many of their tasks (p. 51).

Later in the mission, Walton describes how they modified their routine, which seemed to provide more time for work and leisure (pp. 132, 133). This "popular" change replaced the standard Friday night party that included liquor with a new system in which festivities could occur when people felt like it, resulting in "gay spontaneous parties (p. 133)." He also mentions that they made a great effort to regularly clean their camp and themselves, which was "right (pp. 37, 133)." In other words, these practices helped to lift their spirits.

The crew also participated in a number of other RSPs. Storytelling is mentioned at least once and several references are made to writing or reading letters from home, especially at the beginning of their mission and at the end when Walton is preparing to board the retrieval ship (pp. 27, 88, 121, 186). Diary writing was also important, as was the case for Walton since this ultimately served as the basis for his published memoir. And reading was clearly significant both in the camp at night when the men went to bed, and on their sledging trips when they were stuck in their tents due to storms and heavy winds (pp. 88, 122). Reading material was apparently varied as evidenced by the comment that one night in their tent they read, "*Alice in Wonderland* aloud to each other (p. 118)." The crew also played games quite a bit including "strange paper games," dice, "paper and pencil games," and newly invented games (pp. 88, 119, 120, 125). These were a source of entertainment, especially on their sledging trips, as seen in statements such as how they had "a very hilarious social evening" and these activities "proved popular (pp. 119, 125)."

Moreover, the use of wireless radios, something not available to earlier expeditions, was quite popular. While travelling, Walton reports that his group "listened each night to the B.B.C. news, and the results of the Nuremberg war criminal trials (p. 92)." They also used the radio to receive reports about the progress of another group undertaking a journey on a glacier. In addition, Walton sent a birthday telegram to his father through the assistance of another person he talked to on the radio. He says this is a "very cheerful broadcast" and looks forward to the next one (p. 115). A few days later, several more references are made to cheerful messages between the travelers and their base about various items including the arrival of an American party at another camp close to their own (pp. 117, 118, 120).

The arrival of the American group near their base camp was a fortuitous event for the British expedition. While the American commander was at first reluctant to allow his crewmen to visit the British explorers, he gradually relented as individuals from both groups began to make contact with each other, finding the company of the other party to be a very enjoyable experience. Their interaction involved several kinds of RSPs, one of which used a type of technology not available on expeditions in the nineteenth and earlier twentieth centuries. Witness the author's comment about what became a common event: "The American party had brought with them a cinema projector and generously asked us over twice a week to see their films (p. 141)."

A different activity also stands out. While references to this type of RSP are sometimes found on earlier expeditions, rarely do they match the intensity of Walton's comments, perhaps because the harsh and unforgiving conditions many of those earlier ventures faced made such practices much more difficult to engage in. On several occasions he emphasizes the positive effect their dogs had on the crew describing them as more than "haulage animals" but rather as "companions that always gave a welcome" when the men were in a bad mood (p. 39). "Without this leavening influence of dogs" found through routinely talking, playing, and interacting with them Walton says the expedition could not have "run so happily (p. 39)." He also points out how "dog training trips were a never-ending delight" because they allowed him to get a break from the usual "restrictions of base" and reflect about his experiences (p. 46). He then stresses how they would happily look forward to puppy birth and "behaved like expectant fathers," feeding them after weaning, handling, and caring for them on a daily basis (pp. 48, 49, 139–141). The young huskies he says, "inspire a sense of joy and affection (p. 49)." The importance to the crew of their repeated contact and care for these animals is perhaps best reflected in the author's final comments before leaving the polar station:

> I went ... for the last time to visit my dogs and to say good-bye to them. I think I was more sorry to say farewell to them than to anything or anybody else ... we left [them] ... ready to help other folk just as they helped us. (p. 186)

Finally, one of the, if not the most referred to, ritual involves special collective events and gatherings. During the first part of his memoir, Walton happily mentions how every Friday night they had a party with a dinner and liquor, changed clothes, were sociable, and had entertaining arguments (p. 37). He then describes their celebration of Midwinter's Day which was a "public holiday" where everyone joined to prepare an exceptional meal, drink hot rum punch, sing songs, and dress in their "best clothes," an important day also celebrated later in the mission (pp. 56, 57). Moreover, he mentions Boxing Day which was a holiday for all and "mild celebrations" for Christmas because they are waiting for the return of a sledging party (p. 97). When the party returned, they celebrated Christmas on January 11 by cleaning up and wearing "tidy suits," enjoying a large dinner and enormous cake with rum sauce and reading "special telegrams (pp. 98, 99)." Shortly after that, they are called on by a British ship, the *Trepassey*, and entertain their visitors with cake and spirits (p. 104). Even on one of their land trips, when another sledging party visits them around Christmas, the men gather in a tent to exchange news and jokes, cigars, a bottle of rum, and enjoy the "weirdest of parties (p. 176)." So too, on another trip Walton refers to the men gathering in one tent for storytelling, a change of pace, and perhaps a treat to eat, "social evenings" that were "excellent for morale (p. 121)." And later, when preparing to leave for home they invite some of the retrieval officers to their base to share dinner, two bottles of whiskey, and enjoyable "sociability" and "conversation (p. 184)."

In addition to these comments, a number of telling remarks about collective events are made in the latter part of the mission after the Americans move into a nearby camp. For instance, they invite their new neighbors to their base for supper, serve buttered rum that was "an excellent ice-breaker," and share stories and friendly conversations (p. 134). A few days later, Walton describes a birthday party with a wonderful cake and an evening marked by great "fun" and a "most riotous schoolboy rough-and-tumble (pp. 134, 135)." Soon after that he also states that on Midwinter's Day the Americans entertained them with a "magnificent meal of fried chicken and mushrooms" and so on "and we in return asked them ... to join us for a typical British Christmas meal, followed by a social evening enlivened with buttered rum (p. 139)." Finally, Walton tells us that with the new work routine at the British camp, which eliminated the standard Friday night party they enjoyed a "most happy arrangement" in which "gay spontaneous parties would start at moments when we all felt the need (p. 133)." It was a "happy atmosphere in which we lived (p. 133)."

Interestingly all the references to collective events such as these emphasize sociality between everyone and their resulting high spirits; not one reference is made to events concerned with religious themes. Actually, no references are made to religious rituals, unlike some expeditions where such practices are quite prominent.

In conclusion, the members of this polar mission engaged in a number of moderate and highly ranked RSPs, i.e., they were quite salient, often repeated, generally homologous because they focused on accomplishing their work and promoting the happiness of both individuals and the group as a whole, and they unquestionably had the resources needed for engaging in such activities. Some of these RSPs were collective events infused with positive emotions. Most were reconstituted or recreated ritual behaviors from their pre-expedition lives, e.g., diary writing, games, parties, interacting with their dogs, but a few were of a more novel and humorous nature. Two examples being the "Antarctic Swimming Club," a group made up of inductees who had accidentally fallen into the cold Arctic waters, and a crewman who would whistle for a cab every evening saying if one ever showed up, he knew he was "going round the bend (pp. 98, 177)." The RSPs they engaged in were also fairly diverse in nature, certainly more varied than what is found in a number of other missions. And they included individual and informal social rituals. Some were also of a more formally organized nature such as their work practices, certain celebratory events involving collective meals and entertainment, and the showing of films, although organized practices such as lectures or religious services are not found in this group.

Furthermore, some evidence indicates the presence of at least a quasi-program of RSPs implemented by their leader, Bingham, and members of the group. The captain certainly organized their daily practices and work routines, however, it is less clear whether other RSPs were due to the actions of particular individuals, the crew as a whole, and/or their leader. Lastly, we find that some rituals were dynamic in nature, i.e., they could be modified such as their recreational, party routines in the early and latter parts of the mission and the group initiating interaction and sharing certain collective rituals with the Americans.

All of these RSPs contributed to the team's good morale and spirited group dynamics involving solidarity, cooperativeness, open communication, and positive feelings for each other.

Science in Antarctica: A Mid-Twentieth Century American Expedition

Our final mission also occurred in Antarctica and involves the American base that was nearby the British camp described by Walton. Finn Ronne, an experienced Polar explorer and Commander in the United States Navy, describes the expedition in his memoir *Arctic Conquest: The Story of the Ronne Expedition 1946–1948* (1949). The scientific goals of this operation sponsored by the American Geographical Society ranged from geological, meteorological, and biological research to recording earthquakes, making tidal observations, and exploration. These endeavors were facilitated by the use of a weasel, a mechanized vehicle that could be driven on the snow, and several small planes that were used to photograph the frozen environment resulting in more extensive maps of these areas.

The crew of 23 persons was American except for a Chilean cook. Moreover, the crew included two women, Mrs. Jennie Darlington, a wife of one of the crewmembers, and Ronne's wife, Edith ("Jackie"). They performed various tasks including the former managing their library and the latter writing reports about the mission, preparing newspaper articles, and making records of what occurred to be used in Ronne's memoir.

Despite the more "modern" nature of this and the previous expedition, the crew experienced a considerable number of disruptions and dangers. They included encountering very rough seas during their voyage and finding the abandoned United States base they were going to live in, had been ransacked, which delayed their settlement (pp. 36, 37, 44, 54, 55). Other problems followed such as, in the beginning a shortage of shower water, a crewman falling into the icy seawater and nearly freezing to death before he was rescued, and another crewman falling over 100 feet into a crevasse and being trapped in −25-degree weather for 12 hours before he was saved (pp. 80, 90–92, 131–137). Other troubles involved the usual extreme darkness of winter and blizzards in the summer, an individual nearly losing his big toe which had been badly frozen, a pilot almost being killed when he slipped and fell into the propeller of his plane, a man breaking his collarbone when he was thrown from his dog sled, and a geophysicist not being able to complete a land trip due to troubles with his knees caused by strenuous climbing and the cold (pp. 116, 117, 138–140, 176, 177).

Notwithstanding these problems and the major disruption of one's personal life that participating in a long-term polar expedition entails, we find a large amount of discussion about the positive attitude of the crew and their relations with each other. At the same time, we also find some references to periodic disagreements among crewmembers and the problems of living together in such a confined environment. For instance, Ronne states that breakfast "was not always the most sociable meal of the day" because of "arguments and discussions that arose at this time (p. 95)." He then goes on to make several insightful observations about how in all missions, including this one, differences in personality and the rigorous, stressful conditions of expedition life can create a "severe test" for people leading to "feuds, … quarrels, … [and] disciplinary action," i.e., clashes and conflict (pp. 96–99). He likewise points out that "under these conditions monotony can become an expedition's worst enemy" especially in winter when they must stay inside in close quarters so much of the time (p. 97).

Nevertheless, Ronne emphasizes that in their camp "There were many times when things ran smoothly and everybody was friendly and helpful" and the party fell "into three groups – the ship's crew, the aviation group, and the scientists – each of which more or less kept to itself (pp. 98, 99)." Indeed, he asserts that "compared to some expeditions I have been on … our party was one big happy family (p. 99)." In this regard, he stresses how a successful expedition is one, which can still achieve its objectives despite the "human failings" of crewmembers (p. 99).

In almost the same breath, he then stresses how important it is to have a "program" which keeps everyone "busy, and therefore, reasonably happy (p. 99)." Accordingly, Ronne created a very busy work agenda for everyone involving a variety of tasks such as performing many different scientific chores, working in the machine shop, preparing pemmican for both the dogs and men on the upcoming dog sledges, making trail markers, and readying their planes for flying in the summer months (pp. 82–87, 99–111). He also strategically assigned bunkmates with little in common to reduce long conversations, which would have disrupted their sleep – the purpose being to keep the daily activities strong and the expedition running "as smoothly as possible (p. 94)." All of these work RSPs helped to fill their days, prevent boredom, and give everyone a sense of purpose and accomplishment. The impact of these practices is evident in his observation latter in the mission: "We had so much to do that we paid little attention to holidays and anniversaries," a situation different from some successful expeditions which often celebrated important events (p. 182).

A close reading of his memoir, though, reveals that monotony among the crew was countered through a number of other important pastime activities that were carried out on a regular basis. Ronne informs us that, crewmembers repeatedly engaged in different RSPs including recreational activities such as boxing, Indian wrestling, pressing a hand scale, and other competitive sports (p. 113). He then states that they "read, wrote articles on their specialties for the newspaper syndicate, and gave lectures every afternoon to the rest of the party (p. 113)." Topics for lectures included first aid, meteorology, aerial mapping, an eight-week course in celestial navigation, and a similar course on radio code. Moreover they "had movies three times a week, to which our British friends often came," followed by ice cream, cake, and coffee (p. 113)."

Additional non-work RSPs that occupied their daily lives are also high-lighted in his account. They ranged from an initiation ceremony involving paddling, undesirable haircuts, and being dunked in a tank when they crossed the Equator on their voyage to Antarctica to crewmen card-playing on a non-motion-picture night, young pup-seals being made into camp pets, and an individual nursing eight baby huskies with a milk bottle (pp. 36, 37, 72, photographs after pp. 108, 110). Other references to ritualized activities include a comment about movies often being shown in the camp and several avid fishermen repeatedly engaging in their favorite sport (pp. 80, 182).

Furthermore, several collective events are mentioned including celebrations for holidays. The first gathering with their British neighbors involved a lunch, cocktail party, and "pleasant conversation (p. 75)." Then after success-fully completing much of the work, preparing their camp, Ronne declared a holiday celebrating their accomplishments, which included a "fine dinner (p. 89)" Later on midwinter night, June 22, the Americans celebrated their cooperative program with the British by inviting them to a "blow-out" involving decorations in their bunkhouse and a delicious dinner (p. 120). The British were also invited to a dinner celebrating the Americans' help in successfully rescuing three lost British pilots (p. 162). While Ronne did not allow the use of liquor on the expedition except for special toasts several men celebrated his wife's birthday with a bottle of wine (p. 182). On another occasion a plane dropped a large supply of food to two geologists doing research on a sledging trip, which enabled them to enjoy a feast (p. 241). Finally on Christmas Day, the crew along with five Britons enjoyed a relaxing day at the American camp, decorated the bunkhouse, and ate a "colossal dinner" prepared by their skilled cook (p. 245). These references to social contact and collective events with the British, it should be noted, are similar to the accounts provided by Walton of interaction between the two camps.

In addition to commentary about the crew's RSPs and morale, other statements are made concerning their state of mind and social relations. For instance, in the beginning of the expedition a "flag incident" which involved the British protesting the flying of the American flag on the grounds of land-claiming eventually became a bonding experience between the two camps (pp. 60, 61). Reflecting a shared sense of humor among the American crewmembers "the disappearance of equipment and other unexplained accidents" were routinely blamed on "Expedition Elmer" instead of specific persons (p. 65). Moreover, certain foods contributed to the morale of the crew as witnessed by the several thousand pounds of candy brought for the mission becoming a favored pastime activity with some men consuming a pound or more of sweets a day (pp. 71, 72).

So too, Ronne reports that pranks and joking were common, the result being that "with the hearty and varied diet, the vigorous exercise, and the lack of worries, we kept remarkably healthy (pp. 114, 115)." As the end of the expedition approaches, we also learn that their mood remains high because the author says: "Now we had enough spare time for occasional bridge games and clowning, such as burlesque radio skits (p. 252)." Finally, as both crews prepare to leave at the end of their missions the British leader tells Ronne how pleasant the past year's working and interacting together has been and that this has been "just about the happiest association I ever had in my life (pp. 279, 280)." Ronne's assessment is in turn quite warm and affirmative.

To sum up, this expedition is similar in some respects to the British expedition we previously examined in that the American party regularly engaged in moderate and highly ranked rituals, i.e., practices that were salient, frequently engaged in, homologous in their focusing on work tasks and the group, and supported by ample resources. Work rituals involving scientific tasks, plane flights, dog sledging, camp maintenance, and daily duties were quite dominant due in large part to Ronne's leadership. And other RSPs were also participated in consisting of pastime and recreational activities, some of which involved collective events such as special dinners and watching movies. These ritual practices were essentially reconstituted from their experiences and social world prior to the expedition. As with the British mission the RSPs crewmembers participated in were diverse in nature and included personal rituals, informal group activities, and formally organized practices such as lectures and celebratory events. At the same time, we again find no references to religious activities.

Clearly Ronne instituted a program of work and non-work RSPs and made strategic decisions to enable their enactment and group harmony, while the crew also played a key role when it came to activities such as socializing with the British expedition, reading, card-playing, and sports. The result was that everyone remained active and did not succumb to boredom or suffer from too much stress. These ritual behaviors undoubtedly contributed to the crew's cohesiveness, ability to communicate and work with each other, and usually upbeat mood.

Conclusion

The expeditions examined in this chapter are quite diverse, e.g., in terms of size and composition, duration, polar region, and their objectives. Yet in all of them, different types of rituals are engaged in which are far more significant and far-reaching in their consequences that what occurs in the failed missions examined in Chapter 3. And the morale of the crews is higher and social ties and group dynamics are stronger and more gratifying. This is due to factors such as the leaders often being more experienced and effective due to the different programs of RSPs they promoted, some of the crews being relatively homogeneous in their composition, more heterogeneous crews finding ways to share RSPs and form social bonds, and crewmembers being compatible because of temperament, common social attitudes, and/or previous relationships.

In the next chapter, we shall see that rituals and their effects are even more pronounced, especially in light of the challenges some of these missions faced.

5

HIGHLY SUCCESSFUL EXPEDITIONS

J. David Knottnerus, Kevin Johnson, and James D. Mason

As has been stressed, all expeditions encountered disruptive, challenging conditions, and experienced deritualization that created a major change in the lives of individuals. However, whether crews engaged in reritualization is open to question. Here we turn to undertakings where ritual practices were extremely strong and commonplace, a situation which profoundly impacted the spirits and interpersonal relations among crewmembers.

We will examine four highly successful missions. These polar excursions, like the previous missions, differ in a number of ways including their goals, how well they are known, the size of the enterprises, when and where they occurred, and the crewmembers' backgrounds. In several cases, the success of these missions is especially striking given the hazards they encountered and length of time of the ventures.

The Discovery of the Northwest Passage: The Trials of the M'Clure Expedition

The M'Clure expedition, an English mission that began in January 1850 and ended in 1854, was one of the many efforts to discover what happened to the missing Franklin expedition. The expedition was commissioned by the British Admiralty and was composed of 66 men who were all British except for a German clergyman. Still, while scientific or geographical research were not the goals of the expedition, Robert M'Clure and his crew became the first explorers to discover by sea and sledging the Northwest Passage, a sea passage along the northern coast of the North American continent through the Canadian Arctic from the Atlantic Ocean to the Pacific Ocean. Unfortunately, in doing so the crew became stuck in ice for a large part of the approximately,

DOI: 10.4324/b23044-5

three years they were on their ship the *Investigator*. Although the vessel ulti-mately had to be abandoned, the men were eventually rescued by another British ship, the *Resolute*.

To examine this long-term journey, we employ two key sources. The first account is by Captain Robert M'Clure and is entitled *The Discovery of the North-West Passage* (1857). M'Clure was an experienced seaman who had served at many different grades during his career. The book is edited by Sherard Osborn, a British military officer, who was not a crewman on the expedition. He pre-sents a narrative of the mission derived from the logs and journals of M'Clure. The second source is by Alexander Armstrong, the chief physician/surgeon for the expedition. This work, based on his own records, is entitled, *A Personal Narrative of the Discovery of the North-West Passage* (1857).

Given the nature and duration of this mission, the crew faced numerous threatening, disruptive situations. Many of these conditions are not that different from what we have seen in a number of other missions discussed in this book. M'Clure, for instance, refers to their encountering gales sometimes accompa-nied by snow and ice while sailing to the Arctic, being immersed in fog which impeded their vision of dangerous coasts, running into dangerous ice floes and trying to sail in packed ice, and fearing on many occasions their ship would be crushed by the ice they were trapped in (1857, pp. 29, 38–40, 61, 62, 82–84, 111, 116–120, 201–203, 207–210, 218–220). Other unsettling problems included discovering over 500 pounds of their preserved meat was spoiled, finding at other times that more of their preserved meat had decayed, and later in the journey the men becoming increasingly ill due to scurvy (pp. 123, 132, 133, 253, 273). Of course, there were the dark and extremely cold winters when their ship was incapable of moving. In that regard, M'Clure refers to the "long [dark] night of a polar winter" which lasted for 11 weeks in the first year of their voyage (p. 149).

Armstrong also addresses these issues in his memoir. While much of the chief physician's commentary addresses topics such as the activities of the crew, his account also highlights the difficulties faced by all. For instance, he describes on several occasions the bad storms they experienced and how at one point in their journey the *Investigator* was trapped in ice and nearly crushed causing great anxiety amongst the men (1857, pp. 295, 417–424, 460, 554). He also describes how on sledging trips some men experienced frostbite and a crewman's toes were frozen, gangrene set in, and some of his toes were amputated (pp. 327, 328, 341, 342). And he explains how later in the mission food supplies were reduced resulting in the crew suffering from hunger for many months (pp. 467, 544–546, 573, 574). Ultimately most, if not all, of the men became sick in dif-ferent ways (pp. 514, 531, 532, 555, 568, 571).

Regarding morale and ritual dynamics of the group, we will first examine M'Clure's account of the first two to two and a half years of the expedition, which includes the first two winters, where the ship was frozen in ice. We will then focus on the remainder of the expedition and the men's final winter.

During the first period, both sources provide numerous statements about the high spirits, positive social dynamics, and enactment of many ritual practices by crewmembers. For instance, in the early stages of their sea voyage Captain M'Clure makes several reflective observations about his admiration for the crew, the "devotion and enterprise" of the seamen, their selfless service to their country, and their "enthusiasm (1857, pp. 22, 103, 104)." The editor also tells us that M'Clure was always concerned for the crew's "health and comfort" and they in turn "repaid him with unbounded confidence and hearty exertions (p. 30)." And later, the captain praises the crew's discipline, character, and commitment (pp. 118, 119). Moreover, various references to the morale of the group are made such as the men and officers alike sharing an "enthusiasm," and "loud songs and cheers" as they overcome each difficulty and advance mile by mile on their mission (pp. 61–62).

So too, a number of comments are made about RSPs engaged in by the crew, various special shared events, and their impact on the men's mood and interaction. For example, early in the mission, they meet a group of Eskimos (referred to by the French word "Esquimaux") who celebrate a whale kill with dancing and a large meal, which the crew was invited to and much enjoyed.

In a different vein, various statements are made by the editor suggesting that religion was important for the morale and resolve of both M'Clure and the crew, although less commentary is provided concerning specifics (pp. 107, 108, 141, 220). For instance, when a safe bay is found to winter their ship, M'Clure names it as "Bay of Mercy," a "divine service" is held on their "second Christmas-day," and in his diary he thanks "all-beneficent Providence who has so wonderfully upheld us in our many trials and difficulties (pp. 222, 238, 266)."

Work rituals involving official duties such as maintenance and sailing of the ship shaped much of the men's daily lives. M'Clure's aforementioned praise for the crew and their enterprise center to a large degree upon their being hard, steady workers. Note, various statements to this effect, including how early in the journey they engaged in "much heavy work" re-stowing and examining their provisions, there being a great deal of work to be done at the end of a winter season, all the work they engaged in to house the vessel and create snow embankments to shield her from gales before the coming winters, and cleaning the ship (pp. 32, 123, 164, 261, 262). Work routines played a central role in the lives of all crewmen, even to some degree during the third winter when they became physically impaired.

With that said, evidence indicates that the men participated in a number of other RSPs that were meaningful and rewarding. Sometimes individuals such as the captain, the doctor, and/or others left the vessel to take walks as they did on their second Christmas day (p. 238). We learn that parties of men and officers would sometimes "stroll out" from the ship which helped break the "monotony" and with the return of the sun at the end of one of their winters officers and crewmembers would take longer and longer walks that provided a welcome diversion and a source of "tales" about what they had seen and done (pp. 125, 161, 162, 164). Later, the Captain describes several interesting walks he took into the "interior (pp. 212, 213)." And elsewhere references are made to the "daily excursions of men and officers" and how they would "extend their walks" when increasing daylight allowed them to do so (pp. 216, 244).

Other observations by M'Clure indicate that the crew also engaged in ritualized activities involving "energetic exercise" (p. 244). This was the case even during the third difficult winter when their energies focused on taking "exercise and keep[ing] themselves and the ship clean (p. 262)." Furthermore, M'Clure states that the crew would engage in another type of RSP during the first couple years that was apparently enjoyed by all, organized games, especially "outdoor sports" when conditions permitted it such as "rounders," a British ball game similar to baseball (p. 162).

Another activity that occurred in a more sheltered environment was reading. For instance, M'Clure describes how on a sledging trip he read a story to his companions one night before they fell asleep in their tent (p. 137). Moreover, the Captain's aforementioned reference to "tales" shared by the men about their walks suggests that story telling was an enjoyable social experience for many (p. 162). More generally conversations and debate about different topics provided a type of ritualized activity that the men benefited from as when on Christmas Day in 1850 each crewmember "talked of home" and they "calculated" when those close to them back home would be engaging in various activities such as going to church or having dinner (pp. 151, 152).

We also learn that individuals would sometimes participate in personal or informal projects that were not a part of their official work duties. These personal or in certain cases quasi-official, work rituals could be quite valuable to crewmembers. One such practice mentioned by the captain involved "a winter schoolroom ... which thirty pupils" attended "to learn to read and write (p. 148)."

Another RSP involved different kinds of recreational activities. These more playful practices served to entertain and amuse the crew. For example, oftentimes after the day's work, the lower deck was transformed into a stage on which certain individuals "performed, danced, sang, or recited" for the enjoyment of the men, which resulted in "roars of laughter and light-hearted jokes (p. 123)." A very different source of amusement for the crew involved two ravens who consistently outwitted the ship's dog by getting first to the food scraps thrown out after the men's meals (p. 226).

Interspersed among these practices were occasional special celebratory events that crewmembers participated in. When the Northwest Passage was discovered one of the first things the men did was to create a "bonfire," break into cheer, and have an "extra glass of grog [an alcoholic drink], given them by their leader, which added to their happiness (pp. 141, 142)." Following this discovery came another celebration during their first Christmas that was spent in "cheerfulness and in good fellowship," a huge, delicious meal, and shared recollections and conversation (pp. 151, 152). Then came New Year's Day which was a "happy one" in which the captain and officers shared many delicacies such as mutton (p. 156). And their second Christmas celebration was an all-day affair marked by a religious service, a walk, a special dinner including venison, an "extra allowance of grog," and "dancing, skylarking, and singing [that] were kept up on the lower deck with unflagging spirit, good humour, cheerfulness, and propriety (p. 238)."

Describing the event, M'Clure states that those back home "would scarcely imagine, otherwise, that the crew of a vessel two years upon her own resources in these ice-bound regions, could create such a scene of enjoyment amidst so many gloomy influences (p. 239)." Further accentuating this point,

> several of the petty officers assured him afterwards that, during many years' service in Her Majesty's navy, they had never passed a *happier Christmas*, nor one in which there had been *a feeling of more perfect unanimity and good-will*, – a feeling shared by every seaman and marine in the ship's company. (p. 239) (*Italics mine*)

Lastly, a number of references are made to the men hunting, an activity which contributed to their food supply (p. 187). While oftentimes their hunting trips were unsuccessful, members of the crew including officers repeatedly engaged in this practice (p. 161). The game that was shot was quite varied and included ptarmigans (a northern grouse), deer, and hares (pp. 164, 223, 244, 247–252). And sometimes their efforts were amply rewarded. For instance, on one occasion, we are told "no less than twenty head of deer were hanging up round the ship, yielding a *thousand pounds of meat* (p. 252)." While hunting undoubtedly had a utilitarian value because the meat was always eaten evidence indicates that, at the same time it was a ritual practice that was quite meaningful and rewarding to the men as reflected in the fact that hunters are usually referred to as "sportsmen," how often the crewmen engaged in the activity even though they were not always successful, and the enjoyment they derived from this practice (p. 247).

In sum, crewmembers and officers engaged in a large number of RSPs. Numerous comments are made about these activities, along with high morale and rewarding, enjoyable group interaction. In fact, no references are made to poor morale or disagreeable social relations among the men. And in addition to the commentary just reviewed, other positive statements are made about the high spirits of the crew such as their joking with the aforementioned group of Eskimos they had encountered, cheering at the departure of sledge-parties, cheering at the return of the sun after a long winter, cheering at the break-up of the ice the *Investigator* was stuck in, and so on (pp. 74, 75, 113, 134, 157, 158, 161, 166, 200, 226).

Turning to the last year of the expedition and the third winter conditions significantly deteriorated creating a very dangerous situation. Very likely, if they had not been found and rescued by another British ship, the *Resolute*, in April 1853, many of the men would have died or at the very least become extremely ill. The previous two winters and the summer months, which had been colder than normal, meant that ocean ice prevented the ship from sailing out of the area they were in. Gradually food supplies shrank (pp. 261, 272). The worsening conditions also included developments, such as the lower deck becoming "damp and wet" because they could not "afford more coal to dry it up (p. 272)." Meanwhile, the sick list continued to grow due in part to scurvy resulting in one third of the crew being "placed in the doctor's hands (pp. 272, 273)." As their situation worsened it became increasingly difficult to carry out various tasks including hunting. For these reasons M'Clure developed a plan for two travel parties of 15 men each to set out in the spring of 1853 in the attempt to reach Greenland and inland Canada. The sickest men were to make the journeys; even though Armstrong, the doctor, strenuously objected believing these men would never survive such arduous trips.

Nevertheless, the crew continued to engage in certain RSPs. Work projects such as preparing the ship for winter, cleaning the vessel and themselves, and each of the crew taking turns "being cook or carver for the mess" [the distribution of food rations] were carried out (pp. 261, 263). And if possible, the men took "exercise (p. 262)." Despite sickness and the bitter cold, some were also able to go hunting. Sometimes they were successful as on one occasion when the hunters killed several deer (pp. 272, 273).

Lastly, "two festivals" or celebrations were held, one on October 26, which was the anniversary of the discovery of the Northwest Passage, and the other on "Christmas-day" (p. 264). Extra issues of "food and some wine" were given to all on both occasions (p. 264). For Christmas, the men managed to put together a special meal that included English plum pudding, venison, hare soup, and some musk-ox beef that had been in the ship for over two years (pp. 264, 265). All concerned exhibited "good-will and the determination to be merry, in spite of adverse circumstances (p. 265)." The celebration included songs composed by different individuals, paintings of "past scenes of peril and adventure," and comedic performances in which, at least for awhile, the "sick half forgot their maladies" and everyone "tossed care and anxiety to the winds, and felt thankful for the past and hopeful for the future (p. 265)."

Reflecting the conditions they faced, morale varied during this period. For instance, in November when the winter has set in "those of a very nervous temperament became easily excited and unreasonable (p. 262)." And just before they were rescued, which created jubilation among the men, reference is made to "despondency" on the ship (p. 291). At the same time, M'Clure states that in late August when winter is beginning, we are "ready to combat its rigours as cheerfully as on previous occasions" and when he announces his plans for the travel parties in the coming spring the men "cheerfully received" the announcement (pp. 258, 260). And then there are two festivals that crew made every effort to enjoy. Other commentary, however, suggests that morale was decent but not great and there was an ambivalent quality to it including both positive and negative feelings. For instance, after the festivals "there was not much to make men light-hearted or merry although they were still determined to look as much as possible on the bright side of things (pp. 270, 271)." And then in April 1853 after the first death of a crewman the "moral effect of a death at such a time was distressing" although the captain's speech aimed at increasing morale resulted in the "poor fellows" talking "more cheerfully" and looking "happier (pp. 288, 289)."

In essence, the spirits of the crew were mixed ranging from poor to very good. At the same time, it should be appreciated that no references are made to social conflict or a lack of cooperation and communication among the crew. From all accounts, the men continued to successfully function as a tight-knit group.

Turning to Armstrong's (1857) memoirs and the first part of the expedition we find repeated statements about positive social conditions and the large number of RSPs engaged in by the men. Just like M'Clure, the doctor in the early stages of their journey speaks highly of the men emphasizing their "zeal, activity, and fine seaman-like qualities (p. 4)." A little later he refers to the "fine spirit" and "cheerfulness and readiness" of the group (p. 40). And he stresses how the men are hard workers who exhibit "zeal" and "energy" as when they were in boats towing the ship through ice and there was a great deal of "laughter" and "hearty cheer (pp. 92, 93)."

Numerous comments are made about the morale and interpersonal relations among crewmembers. For example, despite bad weather the men on one occasion were in "high spirits" and laughing as they skated and slid on a pond (pp. 220, 222, 223). Later, Armstrong emphasizes how everyone enjoyed playing games and sports on the ice and taking frequent excursions on land (p. 302). He also notes, how the crew faced a deadly fate due to bad ice with composure, firmness, zeal, cheerfulness, and good conduct (p. 236). Certainly, developments such as their being trapped in a bay due to ice was saddening (pp. 466, 473). And such conditions could create "monotony and [a] dull routine" in their "cheerless environment," which was sometimes "dark and gloomy to a degree (pp. 490, 495)." Yet all other commentary highlights how they tried to be active and keep up their spirits, which at different times involved special collective events and celebrations, which will be later discussed (pp. 296, 486, 487, 491, 492, 495, 497). Such ritualized activities helped keep the men in "good health and spirits (p. 296)."

As already pointed out these activities were quite varied involving for example, hunting which provided "agreeable healthy excitement, interrupted the monotony" and dispelled "winter gloom (p. 485)." Such repeated practices provide, Armstrong says, a "very exciting sport" that is marked by "spirit and energy (pp. 497, 498)." In a different vein, the doctor's praise of one of the petty officers is also telling because he stresses how this person who was correct, trustworthy, and experienced had a great influence on the group. Armstrong's overall assessment is that they had an "excellent crew" whose qualities are perhaps best exemplified by a crewman who became lost while hunting but was able to find his way back to the ship and was still "good-humored and cheerful" about his misadventure (p. 506).

These observations about the morale and social dynamics of the crew are intermixed with numerous remarks about different RSPs. Armstrong refers several times to regular religious practices, particularly the "usual Sunday routine" of "performing Divine Service," stating on one occasion they were "truly thankful for God guiding them on their dangerous trip (pp. 213, 214, 446)." Moreover, formal work rituals are referred to at various times as when he mentions the "routine of labor" and specific tasks the crew engaged in such as preparing the ship for "ice navigation" or working on their sledges at the end of winter in preparation for travels on land (pp. 71, 72, 92, 93, 247, 309). And the previous discussion about morale and social conditions emphasizes how the seamen zealously attended to customary sailing and maintenance duties along with special tasks such as towing the ship when in ice.

The physician also emphasizes how the men enjoyed taking walks. Sometimes this oft repeated activity was done alone or with one or two others such as, when Armstrong describes a walk involving the captain, a Lieutenant, and himself, "solitary rambles" he took, or a hike he and an attendant went on collecting "specimens" and conducting a "geological examination of the land (pp. 305, 333, 527–528)." Other times he describes larger groups of men who would take outdoor excursions oftentimes exploring the natural environment or parties of men who rambled about doing things such as chasing a large group of reindeers they came upon (pp. 395, 412–414, 425, 473).

So too, he calls attention to how the crew would engage in regular exercise. While he and M'Clure do not provide many details about this activity it was an important RSP. Armstrong refers to the "routine of labour and exercise" all engaged in during the Arctic winter (p. 247). The work was light and dealt with maintenance inside the ship while the latter involved at least six hours of outdoor exercise everyday. Specific RSPs mentioned by the physician included organized sports. On one occasion he says that the crew toward the end of one winter in March enjoyed "out-door sports and games on the ice (p. 302)." Such physical activities he stresses greatly contributed to the "maintenance of health" of the crew and helped keep them buoyant, cheerful, and countered depression caused by the long cold winters (p. 247).

A very different RSP involved reading. Armstrong describes how in the evenings a man would read aloud to the others as they performed various tasks (pp. 290, 291). Later, he states that reading was their "principal amusement" and that stories were retold again and again and sometimes changed by the narrators (pp. 491, 492). Actually, a key RSP involved story telling. Armstrong says that they derived great pleasure during their long, solitary nights listening to stories about numerous "incidents of sport" frequently involving hunting (p. 485). Consistent with M'Clure's account conversations were also an important RSP. The doctor reports that the men made an effort to discuss different events, which led the crew to rejoice that another day had passed (pp. 491, 492). He then follows with the observation that they sought "conversation and debate" about any incident such as a young bear who had approached their ship (pp. 492, 493).

Another striking ritual practice the crew participated in involved informal or personal work projects. Just like M'Clure Armstrong refers to a school each evening during the winter for "reading, writing and arithmetic" that was well attended and "agreeable to many (p. 288)." Quite importantly the doctor also emphasizes how many crewmen kept busy in the evenings practicing various trades, e.g., "tailors, bootmakers, knitters: a great variety of needlework (pp. 290, 291)." He tells us that all were self-taught. And as previously mentioned a man would read aloud to everyone at this time. Armstrong says this was how the evenings were spent while the days were occupied in "exercise" and "light duties (pp. 290, 291)." Later, in the expedition he again describes how the men busied themselves in informal work, acting as "tradesmen" such as tailors and bootmakers (p. 490). The range of tasks are further addressed when he says that some did "needlework, knitting, crochet, [and] repairs to caps and mits" and so on (pp. 491, 492). And he praises their excellent work (p. 490). This extensive system of informal work became so important that gun wads were used as an alternative form of currency among crewmembers, which were to be later exchanged for money upon their return to England.

In a different vein M'Clure's account suggests special celebratory events involving positive collective emotions enlivened the crew. Here too Armstrong makes several references to such collective rituals and the impact they had on the group. For instance, he describes how they celebrated landing on a newly discovered territory by planting a flag in the ground, cheering, and placing a scroll in a bottle (p. 209). Later he discusses how they built a cairn (landmark), planted a flag, and cheered when they took possession of a large island (p. 266). He also explains how the crew "celebrated the taking possession of Prince Albert's Land" by being provided with "extra provisions and spirits" and how they "amused themselves" in the evening (p. 264). He then states: "Events such as these tend to maintain men in good health and high spirits from the cheerful excitement they afford (p. 264)."

With the arrival of their first Christmas, they collectively celebrated eating choice foods such as beef and musk ox while there were "many a prayer breathed and hope expressed" that next Christmas would not be in the Arctic (pp. 292, 293). Another collective event involved the men honoring the Queen's birthday with a royal salute involving the firing of guns, cheers, the crew being issued extra "grog," and "hilarity" in the evening (pp. 331, 332). Later, the crew celebrates the anniversary of the discovery of the Northwest Passage with an "allowance of spirits" which had a "hilarious and cheerful influence (pp. 486, 487)." And while their second December was rather "dark and gloomy" they "gladly welcomed" Christmas, were cheerful, and "regaled" themselves with a large buck (deer) for their dinner (p. 495). This shared event was then followed by New Years Day being a "festive one" with "extra provisions" and they're being as cheerful as they could be given the circumstances (p. 497).

Finally, Armstrong makes numerous references to hunting. Actually, the author makes far more references to this RSP than any other activity, so many that it is not possible to discuss all of them. At least thirty-eight comments (on separate pages) are found in his memoir although some pages contain multiple references to hunting so the number is even higher (pp. 169, 275, 276, 292, 297, 298, 304, 309, 313–316, 322, 323, 325, 337, 346, 347, 395, 412, 413, 415, 416, 425, 473, 474, 475, 477, 479–483, 485, 488, 497–499, 504–506, 511, 514, 515, 521, 525, 526, 529–531, 537–539). The type of game hunted and killed was quite varied, e.g., reindeer, ducks, gulls, musk oxen, foxes, hares, bears, seals, and different sorts of birds such as grouse, cranes, and snow geese. Sometimes, one or several persons would go hunting and other times larger parties.

Several comments by the doctor also indicate that hunting was not simply of utilitarian value but that it was a gratifying RSP that was repeatedly engaged in by crewmembers including the officers. Note Armstrong's observation that hunting deer helped compensate for "our want of sports (pp. 298–299)." Later, in the mission he underscores how hunting reindeer provides some "very exciting sport in their pursuit" and the chase is marked by "spirit and energy (pp. 497, 498)." And at another point he elaborates upon the benefits of hunting by stating that these "incidents of sport" were quite "numerous" and they provided much pleasure to the crew as they listened to each other's stories about their adventures during the long, solitary nights (p. 485). He then states that hunting provided "agreeable and healthy excitement, interrupted the monotony," and dispelled the "winter gloom (p. 486)." These pursuits "kept the mind in a state of buoyancy," which countered the "depressing effects" of the dark Arctic winter (p. 486). For these reasons, every day some of the men went hunting (p. 499).

To recap Armstrong's portrayal of life on this expedition is similar in a number of respects to the one provided by M'Clure. The doctor describes, sometimes in greater detail and depth than M'Clure, the numerous RSPs engaged in, their high frequency and salience (high rank), the high morale of the crew, and the cohesiveness, cooperation, and smooth interaction among crewmembers.

In the last part of the expedition, conditions such as the growing sickness of the crew and dwindling food supplies present the men with an increasingly threatening and deadly situation. Nevertheless, Armstrong's account shows that the crew still participated to some degree in certain RSPs. "Uncomfortable" conditions on the "decks" due in part to the cold did not prevent the men from performing the "daily routine" of exercise (p. 548). And work was carried out such as preparing the ship for winter (p. 549). For the crew, the "chief occupation was the chase [hunting]" although, as the men became more debilitated fewer individuals participated in this activity eventually resulting in most of the hunting being done by the officers (pp. 549, 550). Still, references are made to hunting game such as deer, hares, grouse, and wolves although these efforts were not always successful (pp. 545, 546, 551, 554, 555, 556, 557). While the number of persons who went hunting declined the activity quite likely provided grist for stories about hunting and related incidents, a ritual practice that was undoubtedly beneficial to all (p. 551). And as in former years the discovery of the Passage and Christmas were celebrated with the effort to create a very positive effect, a sense of "*éclat* (brilliance)," while giving thanks to God and being grateful no one had died up to that point (p. 552).

Morale as already noted suffered during the last phase of the mission. Armstrong comments several times about how they're still being stuck in the ice, the winter darkness and cold, increasing cases of illness, and lack of food contributed to the men feeling "cheerless and melancholy," experiencing "a feeling of depression," being bitterly disappointed, experiencing "cheerlessness and gloom," and becoming "dispirited (pp. 542, 549)." Not surprisingly the first death of a crewman due to illness (shortly before they were discovered and rescued in April 1853) also had a "depressing influence" on the men; it was a "melancholy occurrence" that created a "lowering cloud of darkness that hung ominously around us (pp. 559, 560)." Nonetheless the crew was still able to celebrate the discovery of the Passage and Christmas, the latter of which created "enjoyment" for the group (p. 552).

After their rescue, Armstrong then offers his candid appraisal of the crew. He says that for 20 months the men did not have a "good meal" and had suffered from pangs of hunger, which they could not appease (p. 773). However, he goes on to praise in no uncertain terms their "noble spirit and patriotic feeling that had animated the Ship's company in the almost super-human exertions, hitherto made under the most severe and trying circumstances" that few have encountered (p. 574). He extols, "their courage, and daring in many eventful scenes" and stresses that as their bodies shrank from "hunger and cold" they exhibited "patience and fortitude when suffering from disease (p. 574)." Such observations suggest that despite their suffering morale still remained moderately strong, although mixed ranging from low to at times high. They remained resolute and motivated. Even when troubled and unhappy in the final stage of the mission they still exhibited a collective determination and dedication to the endeavor they had undertaken and to their fellows. Such an assessment is supported by the fact that no mention is made, as in the case of M'Clure, of any conflict or breakdown of social equanimity and teamwork among the crew.

In closing Armstrong's account is consistent with M'Clure's. Both sources describe how the crew established a routine of activities early on in the voyage, which continued throughout the expedition except for the last phase of the mission when they declined to varying degrees. Members of the expedition engaged in a ritual program, i.e., a large assortment of RSPs, which contributed to social cohesion, cooperation, communication, positive emotions, generally high spirits, and a determined attitude.

Approximately a dozen different kinds of ritual practices are described including formal work duties, exercise, walks, personal work projects, special celebrations, religious practices, storytelling, and hunting. Some of them such as the last one were especially popular. The variety of RSPs is also reflected in the fact that they ranged from personal rituals such as reading and performing "trade" tasks to informal social practices such as storytelling, and in some cases more organized activities including recreational activities involving singing or dancing, games and sports, educational classes, and religious services. Most of these RSPs were reconstituted practices, activities based on rituals that people had engaged in prior to their expedition. Some, however, were of a more novel character involving modifications of ritualized activities such as the crewmen teaching themselves new trade/craft skills including sewing, tailoring, and boot making.

Both memoirs indicate that the rank or dominance of these RSPs was quite high. Numerous commentaries demonstrate that these activities were prominent and central to the crew in their daily lives. These salient practices whether they involved hunting, walks, work, celebrations, or other RSPs were meaningful and usually enjoyed by the men. They were also engaged in frequently. Their high degree of repetitiveness is reflected in the fact that many were practiced weekly, e.g., religious services, if not on a daily basis. They also exhibited a high degree of homologousness in that they complemented each other and often expressed similar themes such as emphasizing the value of the expedition, the crew, group activities, and personal satisfaction and enjoyment. Lastly, both material and non-material, i.e., human, resources were available to the men for enacting these RSPs. Such resources ranged from the ship and the facilities it provided to the equipment needed to engage in activities such as hunting or different sports. Equally important were various human assets such as having the strength and physical/mental capacities which allowed the men to take walks, hunt, work, read, and so on. The importance of the latter type of resource is apparent in the last part of the mission when sickness, cold, and deprivation of food reduced the crew's ability to engage in some of these practices.

So too, everyone participated in various special celebrations. These events allowed for the expression of strong collective emotions such as pleasure and happiness. And in several cases, particularly when they celebrated their discovery of the Northwest Passage, such events engendered a sense of collective pride that further enhanced the feelings of the men about the value of the venture and themselves.

Based upon available evidence from the two sources various factors facilitated the enactment of these RSPs and the success of the expedition ranging from the homogenous background – English – of the crew which meant the rituals they practiced were familiar to most people to the way RSPs were engaged in from the very beginning of the journey thereby creating a foundation for subsequent RSPs. There was also a clear authority structure made up of experienced personnel including the captain. Consequently, the mission was well organized and properly prepared for. And the crew was to all appearances quite experienced, temperamentally sound, and not subject to any overt forms of social prejudice or personal bias, which could have produced social discord.

As to who initiated these different RSPs it is not totally clear because this issue is not explicitly addressed in the two accounts. Be that as it may it seems very likely that the captain initiated and endorsed a number of practices such as regular religious services and the basic routine of work and exercises. At the same time certain activities were enthusiastically taken up if not instigated by crewmembers such as the informal/personal work projects they pursued, hunting, and going on walks. This seems quite likely given the extremely positive assessments offered by both M'Clure and Armstrong about the crew's abilities, hard work, disposition, and dedication to the mission. These observations involve both general appraisals of the men and statements about specific individuals such as the aforementioned remark of Armstrong about the great influence one petty officer had on the crew and M'Clure's comment about the noncommissioned officer of the royal marines who "did good service everywhere" and distinguished himself "in the execution of any service requiring unflagging energy and marked intelligence (1857, p. 244)."

For all of these reasons, we place this venture in the extremely successful category of expeditions with the proviso that due to the decrease of RSPs and morale in the final phase – for understandable reasons – we consider it to be a relatively marginal case.

Heard Island, Antarctica: A Mid-Twentieth Century Australian Expedition

Our next, more recent expedition involves a group of 14 men (Australian as best as we can determine) who were totally supported by the Australian National Antarctic Research Expedition (ANARE) and were taken by ship to Heard Island in the Antarctic in December 1947. There the men lived for 15 months in a station they constructed at Atlas Cove. Our account of this venture comes from the book, *Fourteen Men*, written by Donald Scholes (1951), who was a radio operator on the mission. The purpose of the expedition was to build a weather station, survey and map the island, and conduct different kinds of scientific research including the study of auroral and magnetic observations, tidal waves, cosmic rays, biological research, and geological and glaciological surveys (pp. 2, 5). Half of the group, Scholes states "could lay claim to the scientific status (p. 14)." The positions held by the men ranged from cook, surveyors, medical officer, photographer, and radio operators to meteorologist, geologists, physicists, and engineer.

The men who ranged in age from 21 to 40, were managed by Commander George M. Dixon who was not a part of the mission, and Group Captain Stuart Campbell who was present at the beginning of the expedition but left before the winter to carry out other duties (pp. 5, 82). During their stay, 14 crewmembers ultimately built 20 huts which served different purposes such as providing a mess, toilet, shower, medical facility, different scientific and biological laboratories, and storage of various kinds of equipment and supplies (p. 149).

Scholes provides a vivid description of the island stating that it was "a depressing place ..." with "little beauty in the gaunt grey rocks, the barren flat and grim precipitous coastline (p. 75)." "Dark clouds" he says almost always covered the island in its "sullen harshness (p. 75)." Yet, he also confesses that there was an "almost indefinable loveliness about it" especially on those rare occasions when the clouds parted revealing the "towering" white mountain which "seemed to dominate" their lives (pp. 75, 76).

Without doubt this was a radically different and perilous world, which evoked a number of different feelings for the author and almost certainly his fellow comrades. As in all the cases examined in this study, the crew's normal lives were significantly disrupted resulting in the breakdown or loss of many of the RSPs, which they had previously practiced.

In actual fact, the disruption and deritualization the crew experienced began even before they began their new lives on the island. On their ocean voyage they encountered an extremely bad gale, which prevented the men from sleeping due to the heavy pounding of waves on the ship. Moreover, it nearly blew some men overboard as they struggled to retie some of the deck cargo including oil drums that had blown loose. And at the end of the storm, they had to turn off both engines for repair, which left the ship helpless (pp. 23–29). When they reached Heard Island, they then faced the challenge of landing. Since the ship had to stay in deeper waters, they used different boats and dinghies to reach shore. The rough surf, large swells, hazardous coastline containing rocks and cliffs, and periodic storms made landing very dangerous, sometimes almost overturning the boats and drowning those onboard (pp. 41–43, 58, 66–72).

After successfully landing and moving their cargo on shore, they next had to establish the camp and begin to form outdoor teams to explore the landmass. Construction of the camp required very hard work every day of the week. After making good progress eventually the men "were allowed Sundays free (p. 97)." Scholes tells us that in the first few months they essentially focused on the "immediate problems of food, shelter, and the habits of simple living (p. 114)." In doing so, they faced serious obstacles due to cyclones that continued year-round including in the summer months. The summer storms would range from "great dust storms" to "snow or rain" which made work difficult because one "slid, flipped and sat down in mud pools and streams (p. 98)." And as the year unfolded, the blizzards became more intense occurring two or three times a week (pp. 172, 173, 182–184, 188).

At the same time the group began to form field teams, which, as soon as they could, travelled throughout the island, coasts, and mountain exploring and mapping the landmass and collecting various kinds of research data (pp. 130–138, 143, 144, 151–157). All of this work continued in good and bad weather. These efforts were oftentimes very difficult, uncomfortable, and dangerous (pp. 155–157, 162–164, 167–170, 235). Challenges they faced ranged from dragging a heavy sled packed with a tent and camping gear, research equipment, and food in soft snow and howling storms, travelling in rough terrain including winding lava gullies and sometimes snow-covered quagmires that one could sink in, sometimes barely sleeping at night in their cold, damp bags, to almost falling with their sled into a deep crevasse that was camouflaged by the snow.

But dangers lurked not just on their field trips or this particular expedition. For instance, an individual who went fishing at another bay by himself slipped and fell with his rifle and backpack into the ocean water. The only reason he survived was because his waterproof pants swelled up keeping him floating while he struggled in the waves to reach land. He then had to walk two miles back to the base camp (p. 161). Expeditions on other islands also showed how dangerous conditions were. For instance, while the crew was celebrating one of the men's birthday, they learned by radio of the drowning of an individual who was skiing and fell through the ice into the water (p. 186). And later, they learned by radio of the death of a pilot whose Catalina plane crashed at another island (p. 216).

Notwithstanding these discomforts and hazards, remarkably only a few references are made to the poor morale of the crew. We learn that when one person cleaned his skis late at night and went in and out of the sleeping hut lowering the temperature the skis were "cursed" by the men in their beds (p. 109). Moreover, one comment is made about how the winter blizzards created many sunless days, which had a "depressing" effect (p. 200). And Scholes states that "disputes" were limited to "an exchange of uncomplimentary epithets," although this could be viewed in a positive light since he also adds that "the two disputants ... just provided fun for the rest of the party (p. 216)."

In marked contrast, numerous statements emphasize the positive morale and strong relations involving cooperation, communication, and positive emotions shared by crewmembers. Individual's high spirits are especially evident in their joking and humorous rapport sometimes involving a few persons and other times the entire crew. For instance, a popular activity, which began early during their ocean voyage to the island involved the "line-book ... a flourishing institution" which recorded "tall [exaggerated] stories" crewmembers told each other (p. 21). Another tradition involved, "the old Portuguese ghost," a spirit that was supposed to be the cause of "eerie" occurrences and eventually "everything (p. 106)." When invoked, the entire crew would engage in collectively shared noise making, shrieks, and intense laughter. A different kind of "horseplay" focused on their hair, e.g., haircuts that were not very well done and some not cutting their hair (p. 109). In this regard, Scholes states that the "great steadying factor" which prevented major disputes was the "fact that each one of us was subjected to his own quota of leg-pulling" and being laughed at which made each person "appear ridiculous to his fellows (p. 216)." This included the author, who was made fun of on several occasions (pp. 147, 170, 171).

Different activities and situations fueled the men's joking and positive state of mind. For instance, "Doc," the expedition's physician, was asked by Expedition Headquarters to collect information about the seals on the island including "particulars" about their sexual practices, a request which caused "amusement" among the rest of the group (pp. 164, 165). Meanwhile, the crew created a conversational style unique to the group, an "incomprehensible dialogue" to anyone else that included "amusing noises ... queer grunts and barks" to express their intentions (p. 247). Scholes states the "craze" which became a "habit" was "a harmless safety valve" which "kept us amused (p. 247)." The men also found radio communications to be both entertaining and emotionally fulfilling whether that involved learning about what expeditions on other islands were doing, joking with the other groups, hearing news from Australia, listening to music, or talking to relatives back home (pp. 123, 159).

On a daily basis, the evening hot meal with a hot drink was the "looked-for moment of the day," which "kept their spirits up (p. 137)." The author also refers to the "amusement" provided by outdoor activities like learning to ski although a shortage of skis prevented some from using them during the winter (pp. 160, 161). When conditions were right field trips and hikes could also be uplifting, as in one case where two crewmen "enjoyed one of their best days on the island (p. 259)." Finally, numerous references are made to how the men's morale was raised by watching the many different birds on the island, e.g., they "made life full of interest," they provided the "best laughs" and were "fascinating to watch," and watching them resulted in the author being in "better spirits (pp. 262–264)."

In sum, Scholes' assessment is that while there were times when inevitably "temperaments clashed" morale was quite high and in general "harmony reigned supreme" in their "small island community (p. 114)."

Actually, comments about morale and ritualized practices begin early during their sea voyage. For instance, a formal toast was made conforming with naval custom to Princess Elizabeth and her Royal wedding (p. 18). The initiation of the aforementioned popular "line-book" also began at this time (p. 21). And during a rough sea storm the birthdays of two of their party were celebrated with an "extempore orchestra" being assembled, singing, and drinks (pp. 23, 24).

When the expedition reaches the island, we then find not only many more comments about the crew's spirits but a much larger number of references to the many kinds of RSPs the men engaged in. Before more formally identifying and describing these RSPs one observation is in order. In March, the group held their "first and only athletic meeting near the camp (pp. 126, 127)." The first event was the 100 yards dash. The race was run in a rather chaotic and not too serious manner, after which the rest of the program was canceled. Plainly the crew was capable of choosing which RSPs they did not want to engage in and those they did because they found them rewarding.

As for the types of rituals they performed personal celebrations such as birthday parties continued once they reached the island. They ranged from one person singing happy birthday to another crewman while on a field trip, there always being a "special celebration" with a cake for birthdays at the base camp, and celebrating one man's learning by telegram that he was the father of a newborn son, to a birthday party at the base that was the "best occasion of the winter" because the center of attention "presented his companions with five bottles of home-brew beer (pp. 163, 186, 195, 202)."

Be that as it may references to work rituals are far more prevalent. Reflecting the interests and goals of crewmembers work RSPs are extremely salient and frequently engaged in, i.e., highly ranked. In the first stage of the expedition, a number of statements deal with, for example, the initial landing on the island and construction of the base, building the 20 huts, individuals moving into the huts they had built, beginning field survey/exploration parties, and constructing radio masts and starting the radio station (pp. 75–81, 84, 100, 101, 102–104).

As the group settled in an increasing array of different types of RSPs are referred to. To begin with the men's workdays followed a set routine beginning early in the morning with the cook, duty radio operator, an engineer, and several others getting the camp running (pp. 165, 166). Breakfast was served from 7 a.m. to 8:30 a.m. During the day different tasks were carried out, e.g., weather observations, radio messaging, balloon flights, the geologist doing paperwork in his hut, and so on. Afternoon tea was at 3 p.m. and the evening dinner was from 5 p.m. to 6:30 p.m. Most were free in the evenings to do what they liked. Some were in bed by 8 p.m. while others went to sleep later. Sundays were usually more relaxed for those in the camp.

More specifically different kinds of work are described by Scholes. These RPSs ranged from exploring their frozen environment to providing medical care for the crew. Some of these activities involved more mundane, yet necessary and very important tasks such as accessing their outdoors "food dumps" that were covered by snow, melting snow in a stove to obtain water to drink, cook food with, and take showers, and the "Doc" spending nine months upgrading "Admiralty hut" which served as a medical facility in addition to becoming a "biological laboratory" where he conducted research such as examining animal specimens he had collected (pp. 182, 246).

Others also kept very busy conducting different types of research and scientific tasks such as making daily "meteorological observations" every three hours from 5 a.m. until 11 p.m. even in the worst blizzards along with measuring ozone in the atmosphere (pp. 183, 251, 252). Much of this research required going outdoors, sometimes on daylong hikes and even longer field trips, which meant the men often had to camp out at night. In addition to surveying the island, other investigations were carried out including geological studies of rock formations and collecting specimens, assessing the movements of glaciers, and measuring cosmic radiation while climbing a mountain (pp. 167–170, 179–181, 226, 227, 228–235, 256–259). Of course, any geological or biological specimens collected outdoors meant they had to be carried back to the base camp, which was not always easy to do. Work also continued into the evening even on field trips when "field calculations" were checked (p. 233). Finally, as the expedition approached its end the men were busy in the base camp "collecting their specimens and writing up their book work (p. 260)." Moreover they had to clean up the camp area before their replacements arrived which meant they had to "clear all the garbage and useless trash" that had accumulated over the last year or so (p. 237).

Despite the prominence of these different work rituals their daily lives were still filled with many other RSPs, some more individualized in nature and others quite social, involving both informal practices and more formally organized activities. An exhaustive and systematic analysis of Scholes' account reveals a dozen additional RSPs crewmembers took part in. Not surprisingly several comments are made about diary writing involving how the word "blizzard became a common entry in the expedition diaries," one man writing at night in his diary after being in the field all day, and the same man "sketching out a plan for a home he would build" on another evening in the field (pp. 163, 188, 233). Moreover, the author's account often takes the form of a diary with dates and descriptions of events and conditions on specific days.

A different type of RSP seemingly quite simple in nature brought great enjoyment to the men, i.e., the eating of certain kinds of food. In the evenings the cook prepared "'Kai,' the thick hot chocolate" which along with their hot meal "worked wonders" because it was "the looked-for moment of the day when the hot drink warmed their bodies (p. 137)." We also learn that a "generous allowance of sweets, toffees, chewing gum and smokes [cigarettes] was made to each man (p. 116)." Besides the pleasure derived from consuming these items "chocolates and cigarettes became the island currency" used in an informal exchange system by the crew (p. 116). Also of note were the stories and yarns individuals told each other on a repeated basis. The aforementioned tale of the "ghost of the old Portuguese" as the humorous cause of all sorts of events is a prime example of such an RSP (p. 106). Other remarks are also made about how in the evening the men would tell yarns in both the base camp after their dinner and on field trips in the evening in their tents (pp. 137, 233). And while only one comment is made about another RSP it is quite telling. Scholes describes how some crewmembers displayed a number of talents, which allowed them to engage in personal, informal tasks in their spare time. He describes how one companion created a "first-class wardrobe," another repaired "wrist watches and cameras," a third made "watch glasses," and a fourth person made "shopping bags and beach bags (p. 166)."

Another important ritualized practice for the group involved music. When the crew was constructing the camp, they moved a piano and radiogram, i.e., record player, into the recreation hut. They then had a "celebration" in the evening with "song and music," made possible by a "collection of 150 records (pp. 84, 86)." Later, the author describes how he unintentionally "interrupted a back-stage practice" in which three of the crew were playing a clarionet, piano, and cello for the "camp concerts" they provided (p. 114). And toward the end of the expedition, he says that the "radiogram was grinding out the tunes," i.e., the "popular records" that "had been played so often, we knew every chorus (p. 265)."

Equally popular if not more so was reading. A library was established in the rec room early on in the expedition. Scholes offered his services as a librarian and states that they "had two hundred and fifty assorted volumes and a full set of the *Encyclopedia Britannica*" which was used to settle many disagreements (p. 98). Various statements refer to these resources and their use, e.g., the men at night listening to one of the crewmen reading from a book by a naturalist who visited Heard Island in 1873, a "case of magazines" in the rec hut available to all, the author emphasizing that "Library books were always in demand" and "popular books were passed from hand to hand," the men reading at night during bad storms, and the cook who was given time off on Saturdays and Sundays going to his bunk with food to eat and books to read, especially during blizzards (pp. 105, 124, 166, 188, 192). Reading like music was definitely a very popular RSP for the group.

Another RSP involved recreational practices that were often playful in nature, although in different ways. The rec hut contained an assortment of games such as "chess, draughts [checkers], dice, crib, cards, Chinese checkers, dominoes," and so on (p. 98). Playing games such as cards often occurred in the evenings (p. 188). In fact, the desire to play games even extended beyond the group as seen in their initiating a chess match by radio with a South African expedition on another island (pp. 150, 201). Other recreational activities were also turned to such as an "egg-collecting craze" in which all types of bird eggs were gathered when walking outside and brought back to the camp (pp. 237, 238). In spring, several other activities were tried such as a very ineffectual attempt to shoot "birds of prey" with a bow and arrow (p. 238). More successfully "model boats were floated in the cove (p. 238)." And 100 empty bottles were sealed with written messages in them and placed in the ocean in the hope they would possibly reach New Zealand or other islands (p. 238). Lastly, the previously described fad of making different kinds of noises when interacting with each other can be considered an informal, playful recreational activity (pp. 246, 247).

Another type of recreational RSP occurred outdoors and usually involved different kinds of sport or more traditional pastime pursuits that some people found entertaining. One example would be the aforesaid man who went fishing, then fell into the water and nearly drowned (p. 161). Another time different people took a toboggan ride on a 200-yard slope, which provided "sport and real fun (p. 160)." The most referred to activity, however, is skiing, which occurred at different times and locations (pp. 160, 161, 183, 199, 201, 202, 203). Two of these references refer to Scholes skiing along with his using snowshoes on a trip and taking photos on another outing (pp. 199, 203). These activities, perhaps especially the latter, could also be considered recreational RSPs. Actually, the day he took photos was unequaled in his memory in terms of the "peaceful calm and beauty of the island (p. 203)."

Several other powerful, highly ranked RSPs were also frequently engaged in by crewmembers. Walking was one such ritualized activity. Numerous comments are made about, for instance, short walks on nearby beaches or elsewhere including those taken on Sundays by crewmembers, longer walks, hikes taking most of a day, several day treks, journeys in which people might ski or snowshoe to a location and then walk, and camping and walking (pp. 98, 118, 157, 158, 199, 202, 203, 210, 211, 223, 236, 237, 239, 248, 249, 252–254, 256, 257–259). It is not possible to describe each of these occurrences, but it should be appreciated that some involved Scholes or others who went on walks alone, Scholes sometimes walked with other men, and small groups that Scholes was not a part of going on walks of varying duration exploring and/or simply enjoying the sights, e.g., mountains, glaciers, ice formations, the snow covered terrain, coasts, coves, beaches, ocean, animals, and so on.

Compelling comments are made about a number of these undertakings as in the case of the afore-mentioned skiing and walking outing by Scholes when he took photographs (p. 203). On another walk by himself in September, the author goes on a day-long journey with his camera saying it "was one of my most enjoyable days on the island (pp. 210, 211)." The sights included a small grotto, large stalactites, cliffs, a glacier, cascades created by melting ice, sea leopards, and sea birds. And on another trip, two other men climbed over cliffs and the side of a mountain, discovered a "rookery [breeding place] of ... albatrosses," and sled "five hundred feet down a snowdrift" on a glacier, the result being "both men had enjoyed one of their best days on the island (pp. 257–259)." All in all, the walks taken by the crew were salient, frequently engaged in, generally similar in form and meaning given their focus on personal pleasure, recreation, and sometimes scientific value, and made possible by basic resources such as proper clothing, equipment, and nourishment in addition to having the time for such activities.

Another very popular, highly ranked ritual involved the crew observing, having contact with, feeding, raising, and/or caring for different wild animals on the island (pp. 110, 148, 149, 158, 159, 236, 237, 248, 249, 250, 251, 252–254, 255, 256, 257, 258, 260–264, 267, 269, 270, 271). Such RSPs included a man making a whale bird his pet, the doctor operating on a paddy bird who was injured and caring for other injured birds, the men letting a group of different birds stay in the cookhouse, and one of the crew raising a "brood of penguin chicks (pp. 110, 148, 149, 250)." The assortment of animals taken in and cared for at the base camp also included seals, leading Scholes to refer to all of these animals as our "friends," "visitors, or pets (p. 260)."

He also states that "whenever one felt down in the dumps" due to their isolation he found the "best laughs" came from visiting a "penguin rookery" half a mile from the camp whose "diverting antics" made him return "to camp in better spirits (p. 263)." Reflective of the feelings that developed from the almost continual caring for animals by different crewmembers, Scholes tells us that when one person who raised a penguin chick returned it to the outdoor rookery once it grew into adulthood "it nearly broke Alan's heart (p. 271)." Clearly the crew's interaction with the island's wildlife was constant, extremely prominent in their daily lives, similar in nature, supported by resources such as the food and shelter they provided the animals, and emotionally moving, i.e., a powerful RSP that is referred to in numerous comments and reflective observations.

The crew on this expedition also benefited from more modern forms of technology that were not available to earlier expeditions, i.e., telegrams and especially radio. The use of these technological resources resulted in rituals of communication that all crewmembers highly valued (pp. 96, 102, 103, 117, 119, 123, 146, 147, 150, 159, 177, 191, 201, 173, 174, 235, 254, 261). The desire to communicate with the "outside world" is evident in the efforts of two crewmen who early on constructed a "ham" radio for making contact with different people (p. 96). But much more consequential for all of the men was the radio station that was built to send among other things weather reports to Sydney, receive messages about the progress of other expeditions, the movements of ships assisting them and their landing operations on other islands, and talking with the members of other missions (pp. 102, 103, 119, 146, 147, 173, 174, 191, 261). Moreover, the radio was used for communication between the base camp and those on field trips on Heard Island (pp. 235, 249).

At the same time, radio communication facilitated certain recreational practices while contributing to the socio-emotional health of the group. For instance, radio news about other expeditions could be amusing (pp. 123, 177). As already noted, chess games with other missions were also conducted by radio (pp. 150, 201). And the radio allowed for personal telegrams to be sent from home to expedition members at the base while those on field trips could receive their messages and send messages back in the form of telegrams (pp. 117, 235). Furthermore, the crew at Christmas was able to listen to carols and season greetings from Australia (p. 254). Clearly radio communication played a central role in the daily lives of the men, both in terms of expeditionary operations and personally.

One additional kind of radio communication actually takes us to our last RSP. Several months after the camp was built, "a series of weekly broadcasts to the expedition was made from Melbourne" in which "Fiancée's and next-of-kin relatives" could talk with the men along with "musical requests" being played (p. 159). As the group gathered for the weekly occasion in which news was shared "the Friday broadcasts became the highlight of the week (p. 159)." In many ways these were special collective events in which people could share news, information, memories, and feelings of satisfaction, happiness, and perhaps joy not just with their relatives but also their fellow crewmembers. Of course, other collective events occurred on the mission such as celebrating New Years, the aforementioned festivity involving music and song when they were constructing the camp, birthday parties, and listening to Christmas carols on the radio along with a first-rate dinner (pp. 84, 86, 163, 186, 195, 202, 254).

Besides these shared activities, a new tradition also emerged on Saturday nights wherein all the crew gathered together for "relaxation" and "Music and song, tales of the war days, both often proceeding at the same time" with "Liqueurs, spirits, and cigars available for all (pp. 116, 117, 197, 198)." These collective events were "the occasion of the week" in which people might be, for example, playing cards, mending their clothes, or "laughing and enjoying themselves (p. 116, 117). Quite often the evenings, which could go late into the night involved individuals telling stories, some quite exaggerated, about exploits during World War II. They were "pleasant evenings" that promoted "company and fellowship (pp. 197, 198)." Taken together the Friday and Saturday night gatherings were very special collective events, i.e., RSPs that contributed to social cohesion, vibrant discourse, positive emotional states, and the high morale of the team.

In sum, the crew on this expedition engaged in RSPs that were very salient, frequently engaged in, amply supported by different kinds of resources, and quite similar in that, while different in nature, they emphasized the importance of work, good relations within the team, and the enjoyment and gratification of each individual and the group as a whole. More formally, they possessed a high rank or importance within the group. Consistent with this assessment the crew was very selective about what ritualized practices they participated in. They knew what was important to them and what they wanted to do as seen in the outdoor event involving athletic contests, which ended early and was never conducted again. In that regard, we find no references to any sort of religious services, which suggests this sort of ritual was also not valued by the men.

Examination of the text also shows that crewmembers practiced a wide variety of RSPs. Many different types of rituals defined their everyday lives ranging from work to long walks and special collective events. These practices involved both reconstituted rituals and new rituals or at least quasi-novel rituals. Most RSPs reproduced behaviors they engaged in prior to the expedition, but a number were modified in accord with the unique conditions they faced such as making pets of certain wild animals and recreational activities such as an egg collecting craze or the practice of communicating with a shorthand array of words and noises. The RSPs also ranged from personal/individualized practices such as reading or diary writing to informal social interaction and more formally organized endeavors including the Friday night long-distance radio broadcasts.

The Friday and Saturday night get-togethers were also special collective events that generated positive emotions among the men. While a rigorous analysis of the events is not possible, the evidence we have indicates that in both cases people's attention was focused on specific actions and/or objects, the rate of interaction between everyone occurred at a moderate if not higher level, and an interdependence existed between all whether that involved on Fridays both those present and others in Australia using the radio or individuals and subgroups doing different things while yet creating a special collective mood amongst all on Saturday evenings. Moreover, resources were available both in terms of radio technology, food, drink, music, etc. along with feelings of copresence created by everyone being in the same physical space in the recreation hut that accentuated an awareness of themselves as a group. All of these factors appear to have produced a moderate to moderately high level of shared emotions.

Taken together all of the RSPs created, a very effective program of ritualized practices that filled the days of the expedition team. However, it is very interesting that not one reference is made to a leader initiating RSPs. It seems strange that there would be no commentary of this nature unless this reflected to a significant degree the actual social dynamics operating within the group. Setting aside the possibility that Scholes chose not to discuss the role of leaders in the group it seems that there was no identifiable head, i.e., no clear, formal, centralized leader. Instead, it appears that a program of RSPs naturally developed and persisted among the crew. Stated somewhat differently an organic program of ritualized practices emerged among the men.

Consistent with this interpretation are several statements that show how the crew collectively exercised an informal social constraint or control of each other in order to prevent certain undesirable behaviors and did so because of their commitment to more equalitarian relations within the group. The aforementioned line-book involved a very popular practice of recording all extravagant claims made by individuals, thus making them subject to kidding if not ridicule. Everyone was the victim of this quite effective practice for keeping everyone in his place (p. 21). As previously noted, Scholes also stresses that everyone was subject to leg-pulling and being laughed at which made one "appear ridiculous to his fellows (p. 216)." This too was another effective type of informal social control, which helped prevent disputes and uncalled-for, unproductive claims that might arouse the ire of others. And when pointing out how "No one was exempt" from "week-end cookhouse duties" when the cook had time off, he says, "Ours was a very democratic system (p. 115)." This is an observation that could well apply to the overall social climate of the crew throughout the entire mission.

Within this context it is plausible that many if not all of the RSPs we have described were produced by the crew as a whole. Of course, it is also possible that a subtler, more informal system of leadership existed involving several crewmen. But there are no references to this either.

Putting aside these questions and conjectures concerning leadership there are a number of factors that influenced how successfully the crew carried out its duties and performed the RSPs that they did. To begin with, all evidence suggests that the expedition was well prepared for and very organized from the very beginning. Moreover, the selection of the crew appears to have been carried out in a careful and quite successful manner. The homogeneous background of the crew who were Australian also meant they shared an appreciation for many of the RSPs they participated in. The fact that they could routinely communicate with others by radio and telegram provided another meaningful ritualized practice that earlier ventures did not have unless the latter physically came into contact with other parties, thereby reducing their sense of isolation. Also, their performing various RSPs on the ship at the very beginning of their mission laid the groundwork for subsequent activities on the island.

What is more, the account by Scholes indicates that the crew was temperamentally sound, open minded, respectful, and enjoyed the company of others. These traits may well be due to a careful and thoughtful selection of the crew by the powers who organized the operation. Furthermore, crewmembers were by all counts highly motivated and very serious about the mission's objectives. They recognized the value of the various goals and work tasks of the men, i.e., the scientific projects and essential role played by all those with different skills who supported the mission such as radio operators, the physician, and the cook. Notwithstanding the inevitable differences and frustrations that develop in an environment like this, all individuals and subgroups worked together. The men cooperated, communicated, played, and joked with each other and most importantly, as best as we can tell, decided what ritual practices they would participate in.

The result was an extremely successful polar expedition.

Imperial Trans-Antarctic Expedition: Shackleton's Challenge

The next mission we shall investigate is famous in the annals of human exploration and command and has been chronicled in various documentaries and movies. Indeed, some consider it to be the definitive example of strong leadership and fortitude in the face of extreme adversity. The English mission, spearheaded by Sir Ernest Shackleton, is known as the "Imperial Trans-Antarctic Expedition (1914–1917)" or "Endurance Expedition" which was the name of the ship. To put this journey into its proper context, many people believed that after Roald Amundsen had reached and successfully returned from the South Pole in 1911 the next, great goal of polar exploration was making the first land crossing of Antarctica. This was Shackleton's objective. While the mission involved scientific investigation, the greatest emphasis was on the feat of traversing Antarctica.

Having already participated in several polar expeditions, the experienced explorer began by soliciting financial support from essentially wealthy donors. After acquiring the needed financial backing, he advertised (more than 5000 persons applied) and recruited the members of the crew who were all English except for one American. The group of 28 men was composed of an assortment of individuals including officers, engineers, scientists (two of the six being surgeons), and various crewmen. About a half dozen or so of the crew, many of whom Shackleton knew, had previously participated in different polar missions. Probably because they were experienced explorers Shackleton chose a number of these persons to hold command positions on this new undertaking.

We utilize two key sources for describing their journey. The first work is a compilation authored by Alfred Lansing, entitled *Endurance: Shackleton's Incredible Voyage* (1959). Lansing draws upon an assortment of sources to describe the expedition such as the diaries of all crewmembers who kept them, long interviews with most of the surviving members of the mission, letters with questions for the men, other researchers, and the Scott Polar Research Institute. A second work was also used for this study, which provided much important information and personal insights. It is Ernest Shackleton's own account, *South: The Story of Shackleton's Last Expedition 1914–1917* ([1919] 1999).

Because the expedition involved a number of developments, most of which were not planned, we will first provide a chronology of these events and many of the disruptive conditions endured by the crew. *Endurance* sailed from England in August 1914. After reaching Buenos Aires and then South Georgia Island in the South Atlantic the ship set sail for Antarctica in early December. Within a couple of days pack ice was encountered which led to delays in their journey. Then in mid-January the vessel was trapped in ice despite determined efforts to free her. The *Endurance* could only drift being the captive of the ice through the winter months. With time the pressure of the ice on the ship grew putting the vessel at increasing risk. Eventually the wood beams began to break which resulted in the *Endurance* being crushed on October 27, 1915, and the crew abandoning the ship to live on the ice floes. The goal of crossing the Antarctic was abandoned and replaced by the effort to survive.

Various supplies and three lifeboats were saved before the ship finally sank on November 21. During this period and in the months to follow, two camps – Ocean Camp and Patience Camp – were established and several marches were attempted but then halted due to poor conditions such as rough terrain and the failed attempt to haul their boats in very soft snow. Meanwhile food supplies shrank. Seal meat was relied on more and more and eventually all the dogs were shot due to their needing too much of the remaining food and because they became food for the men. Thus, the crew drifted on the ice until the inevitable splitting of the floe they were on occurred on April 8, 1916. The next day, the group departed in their three boats hoping to reach a distant island.

Their treacherous journey took place in the midst of floating ice, in extreme cold as low a −20 degrees Fahrenheit while being wetted by icy seawater, and with limited food to eat. On April 15 they made landfall on Elephant Island, a barren, unpopulated, and isolated landmass, which contained a stretch of land where they could camp. But to remain there would only end in disaster for the group. Shackleton decided they would prepare one of the lifeboats, *James Caird*, for an 800-mile journey to South Georgia Island where there was a whaling station. He selected five men to accompany him in the twenty-two-and-a-half-foot boat. Frank Wild, the second-in-command, was left in charge of the 22 men remaining on Elephant Island. On their boat trip, the six men encountered a gale, extremely high waves, and icy water, which covered the *James Caird* with ice. After 16 exhausting days, they reached South Georgia Island and then rested for several days. They had landed on the southern part of the Island, but the only people were at the whaling station on the northern side. So, Shackleton and two others hiked for 36 hours without a map before finally reaching the whaling camp. A boat was sent to retrieve the other three men of the six-man group the following day.

Four attempts were then made to rescue the rest of the crew trapped on Elephant Island because heavy ice floes prevented the first three rescue ships from reaching them. Finally, after approximately four months, on August 30, 1916, all the crewmen were saved. During their time on the Island the men suffered from a number of health problems, food became increasingly scarce, they were subject to the vagaries of the weather, and uncertainty grew over whether they would be rescued. Fortunately, they were able to turn their two boats upside down and place them on a bed of rocks, which provided them rudimentary protection.

Clearly expedition members encountered a host of disruptive, life-threatening conditions during their tortuous journey. Many of these we've already mentioned. They included severe blizzards encountered by the *Endurance*, months of stress created by the increasing pressure of the ice *Endurance* was trapped in, the painful necessity of killing their dogs, and the fear of being eaten by killer whales while on the ice floes, the extreme cold, and seasickness while sailing to Elephant Island in their small boats. Moreover, during the six-man *Cairn* boat trip, Shackleton tells us how the men struggled to survive in "heaving waters" and never ending "gales" where they were repeatedly sailing between the crests of huge waves and the valleys between them, all the time attempting to rest in soaked freezing clothes and sleeping bags (pp. 54, 109, 131, 135–138, 167–169). On Elephant Island, 22 crewmen faced their own challenges such as the need to amputate a man's foot with limited medical supplies (Lansing 1959: 205–207). The group's home was a spit of land of just one or two acres – "rough, bleak, and inhospitable" – stuck between the sea and a "snow slope running up to a glacier" that was subjected to alternating "cold, damp, misty weather" and "blizzards (Shackleton [1919] 1999: 152, 234)."

Not surprisingly the men's morale sometimes suffered due to the harsh, perilous situation they were in, the long period of time they were alone, their uncertain fate, and the physical and mental toll these conditions took on them. Nevertheless, a thorough examination of the evidence reveals many references to high morale, the happiness of the men, good relations among crewmembers, and their ability to work together as a team. Moreover, many comments are made about individuals engaging in RSPs through the different phases of their venture and the positive impact these rituals had on everyone. Actually, from the very beginning, while on the *Endurance*, the crew engaged in and benefited from an array of rituals. These RSPs and the harmony they engendered helped lay the groundwork for the high morale and group cohesion that was evident throughout their journey.

More precisely, numerous remarks emphasize how different RSPs were engaged in, often on a daily basis on the *Endurance*, rituals that were unmistakably facilitated by Shackleton. A key ritual involved work practices that he made sure the crew followed including "readying the ship for the long winter's night ahead" while stuck in the ice and permanent dog teams which trained "regularly" and sometimes sledged seals that had been hunted and killed back to the ship to be used for food (Lansing 1959: 34, 36–38). The creation of dog teams for different work activities also provided the basis for recreational RSPs involving competitions between dog teams and very spirited dog derbies (pp. 44, 50, 51). Other competitive outdoor activities among the men involved "hockey games on the ice (p. 50)."

A large number of indoor ritual activities on the ship also shaped the lives of crewmembers. For instance, their first Christmas was celebrated with an "excellent dinner" and a "songfest" while people's birthday was celebrated with a special cake prepared by the cook (pp. 27–28, 40). Actually, there was a "regular series of social occasions" which included singing at different times, rolling dice for fun in the evening, everyone being issued grog, an alcoholic drink, followed by a toast every Saturday night, and songs being played on a "hand-crank phonograph" on Sunday evenings (pp. 42–43). Moreover, a "special celebration on Midwinter's Day June 22" was held in which the boat was decorated, a stage for "festivities" was prepared, and comedic presentations, recitations, and songs were greatly enjoyed by all, followed by a late supper, toast, and singing of "God Save the King (pp. 42–45)." The expedition photographer also delivered a lecture once a month on places in the world he had visited (p. 43). And then there were the evening activities of eating their meals, reading, playing cards, and writing in their diaries (p. 43).

Shackleton's memoir ([1919] 1999) also makes numerous references to similar RSPs including work rituals such as the biologist dredging the sea for specimens, photos being taken of different events and their surroundings, training the dog sledging teams, ship duties ranging from measuring the depth of the sea to trying to clear ice and snow from around the ship before it became completely stuck, and all the scientific staff carrying out a variety of tasks such as taking meteorological measurements and searching for geological samples (pp. 24, 25, 26, 31, 36, 39, 40–42, 44, 51).

Furthermore, he describes many different social and recreational RSPs including a crewman entertaining others with his banjo, a "vigorously contested game of football" on the floe by the ship, celebrating their first Christmas with grog, an excellent dinner, a "sing-song," enjoying New Year's Day with well wishes and noise making, and playing hockey and football on the floe after their ship becomes stuck in ice (pp. 12, 13, 15, 17, 36). Like Lansing, he also comments several times on dog team races such as their "Antarctic Derby" that was marked by much betting and cheering and another race a few days after the "Derby (pp. 52, 53)." Finally other social activities are highlighted such as singing patriotic songs on "Empire Day," having during the darkness of winter a very cheerful and noisy concert while noting how they "refused to abandon their customary cheerfulness," and celebrating Midwinter's Day as a holiday with a superb dinner, speeches, songs, and toasts (pp. 46, 47, 50, 53).

These and additional comments demonstrate that the crew's morale was quite high and relations between the men were exceedingly strong. As Lansing (1959) notes, the months of "shared experiences" meant that the men "had come to like one another (p. 39)." Later, he points out how it "was remarkable that there were not more cases of friction among the men … But instead of getting on each other's nerves, the entire party seemed to become more close-knit (p. 42)." This assessment is supported by a number of references to tricks, jokes, humor, and pranks among the crew (pp. 39, 40, 41). For instance, two men, including Wild, the second-in-command, "decided to give each other haircuts" but wound up cutting off all of their hair (p. 42). By the next day "Everyone, including Shackleton, had his hair trimmed down to the scalp" all of which led to "many more pranks (p. 42)." All of these shenanigans, which included the expedition leaders, attest to the crew's high collective spirits and positive emotions. Underlying these shared feelings among crewmembers and the different RSPs of the men was the determination of Shackleton to see that everyone remained as active as possible grounded in a "routine of work and play (Shackleton [1919] 1999: 63)."

Although conditions become much worse once they moved to the ice floes, Shackleton promoted the same pattern of activity, which resulted in the crew still exhibiting a relatively high morale and a strong level of solidarity much of the time. While several short marches were attempted, they had to be abandoned due to the difficult conditions they encountered. Most of their time was spent at Ocean and Patience Camps until the ice began to break up and they had to set sail in search of land. The leader's attention to the "details of existence," which in many ways involved the everyday rituals of the crew is quite apparent (p. 80). As Lansing emphasizes, the men became used to "the well-established day-in, day-out routine of camp life (1959 pp. 79–82)."

Each day would begin at 6:30 a.m. when the night watchman heated up their stove and woke the cook who would then prepare breakfast for the men who rose no later than 7:45 a.m. After eating everyone carried out his normal chores such as preparing food to eat, melting ice into water, trying to improve the seaworthiness of the boats, and constructing a boat pump for future use. Most of the remaining crew went hunting while the dog drivers exercised their dog teams. When a seal was killed, usually by hand in order to save their limited ammunition for the revolver, shotgun, and rifle they had, one of the dog teams would carry the carcass back to their camp. Next came lunch followed by more chores such as reinforcing the sledging harnesses, sorting and packing equipment, and further work on the boats. At 5:00 p.m. the dogs were fed and then at 5:30 p.m. the men had supper oftentimes composed of seal meat, hot watered cocoa, and lumps of fried flour mixed with pemmican (dried meat) or something like lentils. In the evening, various activities were engaged in such as card games, e.g., poker or bridge, reading aloud, or talking with each other referred to as "yarning." At 8:30 p.m. came the official curfew although some continued to quietly talk up to 10:00 p.m. Through the night watchmen were posted each of whose duty lasted one hour.

While this summarizes many of the daily RSPs of the crew, additional commentary provides more details. We are told that in the evenings one crewman played his banjo, a number of men turned out to sing, people played cards, two men played poker everyday, nearly everyone became enthusiastic about playing bridge, and many read including at least seven men who read out loud every night (pp. 75, 76, 86, 104, 113). Furthermore, one crewman repeatedly reread the bible, some often prayed, birthdays were toasted, stories were told (new stories being much preferred to the old, repeated tales), and certain individuals smoked their pipes or cigarettes (pp. 110, 111, 131, 132, 141). And some such as Shackleton would go for walks in addition to many writing in their diaries (pp. 122, 123).

So too, Shackleton ([1919] 1999) refers to some of the same RSPs all members of the group engaged in including work rituals such as preparing gear and strengthening the sledges to carry their boats (p. 81). Elsewhere, he says their days were characterized by working, talking, and eating (p. 91). He then states that they passed time reading the small number of books they saved from the ship and later mentions how they would read, sew (unless too cold to do so), and pass time in conversation (pp. 93, 94, 115). He also emphasizes how "seal- and penguin-hunting was our daily occupation" and the 'apathy" and "frustration" of being trapped on the floes for so long was "dispelled" by their daily hunting parties (pp. 93, 108). Finally, Shackleton refers to their keeping Christmas Day and managing to have a "feast" and holding a "special celebration" on Leap Year Day mainly to "cheer the men up (pp. 102, 111)."

While much of this commentary shows that crewmen did cooperate and communicate with each other, and enjoyed these practices, which oftentimes had a salutary effect on their morale, they still suffered because of the perilous, uncertain situation they were in. Lansing (1959) tells us that sometimes there was an "air of tension" and scant conversation among crewmembers, the monotony could wear on them and when cramped together in their tents due to bad weather the men could get on each others' nerves, tempers might flare, some would become "taciturn," and a "sense of mounting desperation" affected them (pp. 89, 117–120). And at certain times such as when they became soaked while sitting in their leaking tents in the cold rain there was a "depression of spirits (p. 128)." Finally certain events, especially the failed attempts to march on the ice floes, were deflating and led to "dismay," "strain and stress," and difficulty remaining "cheerful (pp. 96, 101)."

But as already underscored, their spirits and camaraderie were also very high a good deal of the time. For example, references are made to the "general cheeriness of the men," their adjusting with "little trouble to their new life [on the ice]," how "most of them were quite sincerely happy," and that "they actually had to remind themselves on occasion of their desperate circumstances (pp. 67, 69)." Furthermore, we are told that the group was "jubilant" when they were able to retrieve a large amount of food supplies from the slowing sinking *Endurance* and later how the party exhibited "optimism and good spirits (pp. 71, 76)." And when the *Endurance* finally sank the men still showed "an astonishing optimism" and were "delighted" and displayed "good humor" when their leader made sure they received a "special treat" of fish paste and biscuits for supper (pp. 84, 85). Indeed, as they coped with their situation the author says they "achieved at least a limited kind of contentment (p. 87)." Moreover, Shackleton ([1919] 1999) tells us that after an initial "wave of depression" when their ship sank the crew quickly became "cheery as usual" and "Laugher rang out from the tents (p. 100)." Likewise on one of their marches he says, "all hands were very cheerful" and the break from their monotony raised their "spirits (p. 104)." As already mentioned, he then stresses that the daily hunting parties for food dispelled the "apathy" some individuals felt due to frustration over their situation (p. 108). And near the end of their stay on the ice floes, Shackleton declares that he "was stimulated and cheered by the attitude of the men" and their confidence (p. 121).

Of course, as time wore on the monotony of the situation and the threats they faced on the ice became harder to deal with. But undoubtedly Shackleton's continuing efforts to promote various work, social, and recreational RSPs helped to counter to a great degree these conditions and their adverse effects on the group.

With the breakup of the ice floe the crew then began an extremely dangerous six-day ocean journey in three boats that took them to Elephant Island. As described by Lansing (1959), there was little opportunity to engage in any RSPs on the trip since they were often busy rowing or poling off ice, extremely uncomfortable, very crowded, and sleeping when they could camp on the ice (pp. 140–175). Yet there are a few references to simple, meaningful practices repeatedly engaged in by the men such as some individuals writing in their diaries (pp. 149, 151). Other times the men's response to their freezing agony was to frequently curse everything while one individual sang several songs and choruses over and over (p. 162).

Shackleton ([1919] 1999) makes several comments that expand upon Lansing's observations. He describes how early in their journey they camped one night on a piece of ice, pitched their tents, and had their "blubber [seal fat] -stove burning cheerily;" the men were "well fed and happy" while "snatches of song came to me as I wrote up my log (p. 125)." Soon after the ice they are on cracks and a man falls into the water (who is saved) yet Shackleton says, "with pipes going and a cup of hot milk for each man, we were able to discover some bright spots in our outlook (p. 127)." Further into their trip, he again refers to boiling a pot of milk and the crew smoking their pipes, two simple, pleasing RSPs (pp. 134, 141). He also reports that while the crew is strained and suffering, they "were doing their best to be cheerful, and the prospect of a hot breakfast was inspiriting (p. 131)." And at another point in their journey, he says that their condition is "pitiable" and due to a lack of water they are quite thirsty with "swollen mouths and burning tongues," yet the men "always managed to reply cheerfully" and would jest which "brought a smile to cracked lips (p. 138)."

The upshot is that despite terrible conditions during their journey the group still managed to engage in a small number of simple RSPs, were able to maintain their spirits, and not give up. They persevered and followed Shackleton's lead, which resulted in their making land.

When they reached Elephant Island, Shackleton knew the 28 men could not survive there, so they prepared one of the boats, the *James Caird*, for a lengthy 800 miles journey to Saint Georgia Island. The treacherous 16-day trip exposed the six-man crew to brutal storms and miserable conditions. And there was a very real chance that they would not successfully navigate the boat to the Island, which would have sealed their doom.

Nevertheless, several comments are made about simple rituals and the men's spirits. Lansing (1959) tells us that periodically Shackleton and another crewmember would roll cigarettes and speak about "many things" and the men would at different times drink "hot milk (pp. 220, 239)." We also learn that their morale ranged from low to high. The first two days they were cheerful but that wore away due to their "uninterrupted misery (p. 224)." Later, however, the men are "relaxed," "faintly jovial," and have a "growing feeling of confidence (pp. 236, 237)." In point of fact, much of this is due to the determination and efforts of Shackleton to see that the men "remain cheerful in order to avoid antagonisms (p. 240)." Shackleton ([1919] 1999) tells us that quite importantly they kept a "regular" schedule of meals consisting of breakfast at 8 a.m., lunch at 1 p.m., tea at 5 p.m., and a hot drink of usually milk during the night. The meals were "bright beacons," which made "optimists of us all (p. 171)." Elsewhere he describes how a crewman would always sing while steering, their mugs of hot milk were "bright moments" during their long watches at night, and they "broke into song" during a sailing trip after reaching Saint George Island (pp. 176, 178, 192). Simple RSPs, high morale a good part of the time, and dogged persistence characterized the crew and their leader.

After resting for about nine days on the island, Shackleton and two of the men then hiked non-stop for about a day and a half over mountains, glaciers, valleys, and a frozen waterfall in addition to sliding down the side of a mountain on an improvised sled. On May 20, 1916, they finally reached the Stromness whaling station, which sent a boat the next day to rescue the other three men. Even during this arduous hike with no chance to rest, they still happily engaged in the RSP of singing together. Shackleton states: "Worsley and Crean sang their old songs when the Primus [stove] was going merrily. Laughter was in our hearts, though not on our parched and cracked lips (p. 202)."

Lastly, we turn to the 22 men on Elephant Island led by Frank Wild who remained there for over four months until they were rescued on August 30. Both Lansing and Shackleton, the later of which read the crewmembers' diaries and had conversations with the men, provide a substantial amount of information about their experiences while camped on the small, barren spit of land by the sea. Lansing (1959) tells us that not surprisingly keeping people's morale high was difficult, especially as their fear that they would not be rescued grew, food supplies dwindled, and some experienced health problems such as one man having the toes on one of his feet amputated by the two surgeons because of gangrene (pp. 206, 207). Lansing notes the "gloom" of the men just before their rescue (p. 278). And Shackleton (1999) comments several times about the low spirits of the group. In the beginning, he says that some of the men were becoming demoralized, discouraged, and were focusing on their "discomforts (pp. 155, 156)." He also says that early on their wet clothes and sleeping bags and "physical discomforts were tending to produce acute mental depression (p. 225)." Other than that, though most of the commentary emphasizes the high spirits of the men and their activities.

Wild it should be remembered was an experienced explorer who knew, perhaps due in part to the example set by Shackleton, the importance of keeping the crew active and cheerful. He promoted many RSPs while they were stranded on the island. Lansing (1959) describes how much of their time involved repeated activities such as reading and re-reading the same books and an old newspaper, smoking their pipes, bartering food rations, carrying out work duties such as hunting for penguins, seals, and birds to eat, obtaining ice for water, every man stoking the fire for a whole day, the doctors attending to medical problems, and hiking everyday to the top of a lookout hill (pp. 198, 203, 209, 210, 214, 278). Other RSPs involved more social and recreational practices including certain persons playing the banjo in the evening, reading a cookbook, talking about imaginary meals and recipes, and talking about sweets with one man who had been a pastry cook (pp. 200–202). One special collective celebration was also held on Midwinter's Day (June 22) when they ate a "hearty breakfast," presented a "program of twenty-six acts" (topical verses, barbs), played the banjo, sang, and toasted the sun's return (pp. 208, 209). The "routine of their existence" was filled by practices such as these along with "promenade(s)" on the 80–100 yard long spit and "climbing to the lookout" to scan the skyline for a ship (pp. 209, 213, 278). Finally, diary writing was an extremely common RSP, which is reflected in the scores of excerpts by Lansing from different men's journals.

Shackleton ([1919] 1999) also describes various ritualized practices such as pipe smoking, continually sewing their clothes, reading, and observing "Midwinter's day, the great Polar festival (pp. 153, 229, 236–239)." And he emphasizes that concerts were held every Saturday night with each man singing a song about another crewman accompanied by banjo playing, one of the items Shackleton saved from their ship because he believed it would provide a "mental tonic" for the crew (pp. 239, 240).

These discussions of RSPs are interspersed with various statements about relations between the members of the party and their state of mind. Lansing (1959) stresses that the crew felt "confidant" as seen in different observations about how by discussing food "they kept their spirits up – mostly by building dreams," "there was … an astounding absence of serious antagonisms" considering the situation they were in, and while morale could vary due to, for instance, the weather the men would still "joke" about their frustrations and problems (pp. 195, 202, 203–205). He also notes how one person says in his diary that despite their living in a "dirty" little shelter, cramped all together, and laying next to others suffering from conditions such as a "large discharging abscess – a horrible existence," they were still "pretty happy (p. 212)."

Shackleton ([1919] 1999) in turn emphasizes how the group's "spirits were cheered by Wild's unfailing optimism" and when rescued "all were alive and very cheerful, thanks to Frank Wild (pp. 235, 241)." In a telling paragraph, he discusses the "energy, initiative, and resource" of Wild which kept the party's spirits up, how each person in his diary "speaks with admiration of him," and his "wonderful capabilities of leadership" which meant not simply "telling" but "doing" as much if not more "than the rest (p. 240)." Of course, a major part of what he did involved promoting the different RSPs the men participated in. In the same vein, Shackleton also praises the uplifting influence of another individual named Hussey whose "cheerfulness" and oft-repeated banjo playing countered any "tendency to downheartedness (p. 240)."

In closing, the members of the Shackleton expedition engaged in a wide array of often highly ranked RSPs. Ample evidence demonstrates that these practices were quite salient, repeatedly engaged in, homologous in that they emphasized the importance of work, being of good cheer, and the importance of the group, and were facilitated by a number of resources. Certainly, Shackleton promoted a program of rituals as much as he could throughout the entire mission. Some of these RSPs involved collectively shared special events and strong collective emotions such as dog team racing contests and celebrations. These collective activities were the focus of everyone's attention, called upon the participation of all and their mutual cooperation (i.e., interdependence), and involved different kinds of resources including a heightened sense of co-presence. The program was made up of many different RSPs that were quite diverse in their nature and included individual, informal, and more formally organized actions. Generally, RSPs were reconstituted from their pre-expedition lives although they were tailored to fit the unique nature of the mission and the extremely serious conditions they faced, e.g., sporting events and contests on the ice floes, conversing about food, recipes, and favorite dishes, and creating work/recreational rituals in the form of hunting parties to find food.

Of course, some situations made the enactment of rituals far more difficult if not nearly impossible to carry out such as the two boat trips that occurred in horrible conditions, the final hike over Saint George Island, and to a certain extent the 22 men's stay on the barren, small spit of land on Elephant Island. Nevertheless, the RSPs crewmembers participated in were still meaningful and effective to a significant degree even when they involved simple practices such as having a periodic hot drink, small meal, or just singing.

Furthermore, a number of factors contributed to the enactment of rituals and overall success of the expedition. They ranged from the homogenous nature of the crew who were English and shared the same cultural background and appreciation for certain kinds of practices to the clear authority structure composed of leaders who the crew respected, liked, and were committed to. So too, the top authority figures, Shackleton and Wild, were unified which meant there was an absence of conflict between them as was the case in certain other expeditions. And they were experienced polar explorers. Shackleton also controlled the selection of crewmembers and exhibited a very keen, intuitive ability to pick individuals for the expedition. While differences certainly existed among the men, they were by all accounts highly motivated, hard workers, lacking in any significant forms of bias, and respectful of their fellows. Additionally, the expedition was well prepared for and well organized, which attests to Shackleton's planning and ability to clearly communicate with others through all phases of the venture.

Of course, Shackleton's effectiveness as a leader was a crucial factor that contributed in many ways to the mental state of the men. As already discussed, his ability to lead is especially evident in his cultivating a program of dominant or highly ranked RSPs that were appealing to everyone as soon as their journey began. The establishment of an array of RSPs on the *Endurance* (and then on the ice floes) laid the foundation for subsequent ritual practices, strong social ties, and high morale among the crew.

As to why he did so Lansing (1959: 73) suggests that Shackleton had a "dread of losing control of the situation" that was partly due to a "consuming sense of responsibility." For this reason, he was willing to do almost whatever it took to "keep the party close-knit and under his control" including watching out for "potential troublemakers who might nibble away" at the solidarity of the group (p. 73). This would also partly account for his constant attention to keeping the men busy by working, playing, and engaging in other activities. From our perspective, he was determined to see that everyone participated as much as possible in numerous RSPs to ensure the high morale and unity of the crew. Of course, Shackleton was a very good leader for other reasons such as when they had to abandon the *Endurance* and discard many of their possessions before they took to their sledges on the ice, he set the example for the crew by first throwing into the snow his gold cigarette case and gold coins (pp. 64–65). His sound judgment is also evident in the decision to let the men keep their diaries and ordering the crewman Hussey to take his 12-pound banjo with him. Shackleton clearly understood how important these items would be for keeping the group active, motivated, and in good spirits during the difficult days ahead (p. 65).

In saying this, we should at the same time remember that the crewmembers often willingly if not enthusiastically engaged in and sometimes initiated different activities whether that involved reading, banjo playing, dog team racing, hunting, concerts, singing, and so on. Still, the extremely effective leadership of Shackleton and Wild on Elephant Island who may not be appreciated as much as he should be, stand out and played the crucial role in promoting numerous RSPs, high morale, and the determination of the crew during the different stages of this harrowing venture.

A Late Nineteenth Century Arctic Expedition: Nansen and the Fram

Our last expedition is another well-known undertaking spearheaded by Fridtjof Nansen. The mission is often referred to as the "Fram expedition," which was the name of the ship used for this Arctic venture (1893–1896). Nansen, an experienced explorer, and the crew of 12 men were Norwegian except for one individual who was Swedish. Their goal was to conduct scientific observations and research, mapping, and if possible, reach the North Pole. Despite employing a new and quite different approach for exploring the

Arctic, Nansen was able to secure funding for the journey from the Parliament of Norway and various private donors including businesses, professional groups, and individuals.

Building upon a model of transpolar drift developed by the meteorologist, Henrik Mohn, Nansen reasoned that a ship could allow itself to be frozen in Arctic ice and then be carried toward to the North Pole. He commissioned the preeminent shipbuilder, Colin Archer, to design, and build the vessel, which had a very rounded hull and remarkable strength due to the size of the wooden beams and structural reinforcements. After the ship was built and all preparations were completed the Fram sailed to the eastern Arctic Ocean and the New Siberian Islands. Once frozen into the ice pack the vessel then drifted in a westerly direction, passed south of the North Pole, and eventually entered the open sea near Spitsbergen, a land mass northwest of Norway. Helping Nansen to do all of this was the second-in-command of the mission and Captain of the Fram, Otto Sverdrup who had accompanied Nansen on a previous expedition in Greenland. It is worth noting that while some approved of Nansen's strategy others, including a number of experienced Arctic explorers, were very critical of his plan and predicted the expedition would fail. Fortunately for Nansen and the crew their prediction proved to be wrong.

Our source for investigating this expedition is the account written by Nansen *Farthest North* (1898). This is a large book, almost 700 pages long, which is written in relatively small print. It provides a great deal of detail about the mission, the crew, and the author's activities and feelings about the enterprise. It is also large because his memoir discusses three key aspects of the expedition. One portion of Nansen's account describes the Fram's trip to the ice pack and the drift of the ship after it is frozen into the ice. This is the part of the journey – June 1893 to March 1895 – which probably stands out the most in people's minds and receives the most attention. The second aspect of the expedition deals with the two-man journey taken by Nansen and Hjalmar Johansen – March 1895 to June 1896 – in which they attempt to reach the North Pole employing both dog sledges and kayaks. Nansen decided it was necessary to do this because the drift of the ship was not going to get them close enough to the Pole. While the two did not reach the Pole, they set a record for travelling further north than any before them. Sverdrup focuses on the third part of the mission – March 1895 to August 1896 – in a shorter account about the rest of the drifting vessel's journey in the ice after Nansen leaves. His report constitutes the appendix of *Farthest North*.

Our study focuses on all three facets of the expedition. These accounts of the voyage are rich with information about the daily lives of the crew. Actually, there is so much descriptive material we are limited as to how much we can quote. Nevertheless, as in the previous missions we examine all aspects of the expedition through a systematic study of comments about the disruptions and deritualization experienced by the group, the negative and/or positive morale of crewmembers, and the different ritual practices they engaged in, along with reflective remarks by the leaders, Nansen or Sverdrup, about the venture.

Before doing so, we would make one observation about the activities of the crew. The first portion of the expedition, which was the longest of the three parts, helped create a foundation or precedent for some of the RSPs that were practiced in the other two segments. At the same time certain RSPs were abandoned or modified given the conditions and/or changes in leadership and composition of the groups in the other segments of the mission.

To begin with the expedition as in many other voyages encountered a number of disruptions and potential breakdowns of crewmembers' usual daily lives, i.e., deritualization. This was the case on the Fram during the first part of their journey (pp. 75, 109, 140, 157, 158, 169–174, 254, 255, 286, 326–331, 332, 333), the two-man trip by Nansen and Johansen (pp. 374, 375–380, 384, 388, 450, 522–525), and the remainder of the Fram's voyage captained by Sverdrup (pp. 610, 621, 651). Of course, the crew's isolation and separation from family, friends, and country for approximately three years made this a very difficult and stressful situation that was compounded by their extreme dependence on nature whether that involved the drifting ice their vessel was in which sometimes would flow very slowly, not at all, or in southerly direction, or the rugged terrain the two men were trying to traverse with sledge dogs and kayaks. For instance, under the command of both Nansen and Sverdrup the periodic slow or non-existent drift of the ice and ship led at times to doubt about the fate of the mission, frustration, impatience, and monotony among the crew.

Other kinds of disruptive events involved extremely serious threats to crewmembers' lives and sense of security. Of course, one concern was the possibility that the ice could crush the Fram. Actually, at different times the pressure of the ice was so great it made loud, crackling sounds, which on at least one occasion made the crew disembark and wait to see if the ice would destroy the ship. Other disturbing factors included the cold and inclement weather, the physician becoming temporarily snow blind, a puppy being killed by machinery on the ship, and a bear coming on board the vessel and killing three dogs.

The two-man journey confronted both similar and different problems. As they dog sledged every day, Nansen and Johansen encountered extremely cold and sometimes wet conditions and at different times great difficulty sledging over rough, icy terrain. Ultimately worsening conditions such as wide-spread ice ridges, cracks, ice blocks and rubble became so bad Nansen decided they would not be able to reach the North Pole, which then led to a prolonged effort to find a more southern body of land where they would hopefully make contact with other humans. Sometimes food ran short on the return trip. Moreover, from the very beginning of their journey the two men brutally drove their sledge dogs. While this bothered Nansen a great deal, he felt it was necessary if they were to succeed. Another distressing practice involved killing their dogs when they became too weak to continue, dismembering them, and then feeding them to the remaining dogs. Eventually the entire group of over 20 dogs was done away with.

Finally on their return trip when they reached land, i.e., an island, they were forced to build a hut and stay in it during the rapidly approaching, harsh winter. Their stay lasted around eight months and was quite challenging as they had very little to do given the weather and their small shelter. Fortunately, after winter ended the men were able to continue kayaking and despite being attacked by walruses were able to reach another body of land where they met the English explorer, Frederik Jackson, who helped them return to Norway.

As is evident in the preceding discussion morale suffered at different times. On the Fram, the main reason had to do with the drift of the ice and ship. When the drift was slow or absent, people could become frustrated, gloomy, and/or bad tempered. Nansen mentions this in the first phase of the expedition (pp. 109, 154, 208, 211, 212, 219, 220, 226, 235, 238, 255–258) and Sverdrup also refers several times in the third part of the mission to the slow movement of the ship negatively affecting the spirits of the men (pp. 610, 621, 651). When the speed of the ice flow picks up, however, the mood always seems to significantly improve. This was certainly the case for Nansen. In this regard, it should be appreciated that his apprehension is a very personal sentiment expressed in his diary. A great amount of evidence indicates that among the crew he was a very confident, strong, and effective leader whatever the conditions might have been.

Lastly, as has been noted, certain aspects of the two-man journey were difficult for Nansen and Johansen, especially the cold and sometimes wet conditions, their brutal handling of the dogs, and the monotony of their lives in the winter hut (pp. 374, 375–380, 522–525).

Yet, despite the disruptions and deritualization experienced by the crew and references to lowered morale, we find that comments about positive morale are especially striking both in terms of their meaning and large number. Nansen repeatedly refers in different ways to the high morale of the group while on the Fram (pp. 123, 126, 127, 132–135, 140, 150, 154, 157, 158, 178, 179, 182, 183, 186, 190, 191, 193, 194, 206, 211, 221, 228, 238, 239, 248, 264, 272–277, 279, 281–283, 287--290, 319, 320, 324, 328, 332, 333, 341, 350, 351, 353, 354, 370–373). Also, some pages contain several distinct references to high morale (pp. 132, 183, 206, 319, 320).

In brief, Nansen's comments about the crew range from statements about their joking, feeling safe, eating exceedingly well, being brave, satisfied, happy, their high spirits, feeling well and friendly, and having a sense of equality to celebrating for reasons such as Christmas or birthdays. He even tells us that several times when the ice was placing great pressure on the ship their morale was still high and they enjoyed themselves. And at one party they read and laughed about the objections others had made about their upcoming expedition. Finally, he makes the rather remarkable statement that while people at home are pitying them over all their hardships, their "compassion would cool" if they could "hear the merriment that goes on and see all our comforts and good cheer (pp. 182–183)."

On the two-man mission, Nansen remarks numerous times about their good spirits and how well they communicate with and support each other (pp. 376, 386, 390, 392, 402, 408, 416, 434, 450, 454–456, 462–464, 470, 471, 484, 530, 531, 537, 539). Again, separate references are made on certain pages (pp. 392, 455, 462). For instance, Nansen comments about how their supper is the best and most enjoyable event in their daily lives, Johansen is a brave, determined individual, a friend, unselfish, and takes care of him, they are in good spirits, the two do not quarrel, they have good, long talks, they celebrate various occasions such as Easter, Christmas, and the weather, and they "were fairly well off" and "our spirits were good the whole time" even when stuck in their hut during the long winter (p. 539).

Sverdrup also discusses the positive morale of the crew on the Fram (pp. 606, 610, 614, 615, 617, 621, 633, 644, 651). Sverdrup's comments range from his requiring the crew to make daily snow-shoe trips which they are happy about, having several festivals for Norway Memorial Day, their celebrating the construction of an ice hut/observatory, enjoying and benefiting from hunting to affirming that "we were all sound and healthy, and had learned to stick closer to one another for better or worse (p. 621)."

Intertwined with these and other references to high morale and positive social conditions are many remarks about the various RSPs crewmembers engaged in. Not surprisingly a major type of practice that defined their lives was work rituals involving all the tasks needed to keep the Fram fully operational while attending to the needs of the crew, plus scientific activities involving different kinds of research and data collection. Nansen provides ample discussion of these practices on the ship (pp. 73, 84, 121, 122–128, 191, 192, 227, 228, 233, 243–246, 258–263, 284, 307–311). They range from cleaning the boiler, preparing the ship for winter, making their own boots, working in the engine room, forging tools, repairing their organ, and preparing for the trip Nansen and Johansen were to take to collecting fossils, the botanist collecting plants early in the trip, routinely taking meteorological, astronomical, and magnetic observations, checking sea depths and temperatures on a regular basis, observing animal life, the doctor conducting medical tests, collecting specimens of ice, sea-weed, and so on, and studying with a microscope plant and animal life.

Nansen's description of work rituals on the two-man trip is more limited in nature (pp. 375–380, 384, 454, 455, 501–520, 553–570). Their efforts largely center on driving the dog sledges or pulling them along with the loaded kayaks after the dogs were gone plus other related activities such as feeding the dogs and preparing the men's dinner and breakfast, kayaking, and constructing their winter hut. Sverdrup's account of life on the Fram after Nansen leaves is more consistent with the latter's discussion (pp. 606, 613, 622, 623, 633). He emphasizes how they did chores such as constructed more snow-shoes, worked on their sledges and kayaks so they could be used immediately if the need arose, built an ice hut for making magnetic observations, continued all the scientific activities described by Nansen, and led a life divided into exercise, rest, and work.

As important as work rituals are, however, many other RSPs contributed to the morale and solidarity of crewmembers. While some of these ritual enactments are referred to more than others, we will identify all of the ritualized behaviors that are discussed. One RSP, which is mentioned three times by Nansen deals with different individuals on the Fram telling stories and yarns (pp. 70, 186, 201). Another related practice that receives relatively more attention involves conversations (Sverdrup does not mention this practice). Nansen refers several times to chats on the Fram among the crew or between him and another crewmember (pp. 140, 211, 266, 272, 273). And on the two-man trip, Nansen describes how Johansen and he had long talks in their hut about, for instance, what they would do next winter to make up for what they had missed, how far the Fram had drifted, whether she would reach Norway before them, and all the good things that were comforting to them like books, different foods, clean clothes, or a Turkish bath (pp. 458, 535, 537). In this setting talking for hours about such topics was extremely rewarding for both men.

Actually, a wide variety of RSPs are participated in which were valued by different crewmembers. Diary writing was certainly very important for at least some. Nansen refers to his diary several times and points out that writing in it removes the despondency he sometimes has about the slow drift of the ship (pp. 99, 101, 144, 190, 197, 204, 221). On the sledge trip he also tells us that the two wrote in their diaries in the morning after breakfast and in the winter hut although there was little to write about (pp. 377, 522–525). Sverdrup doesn't explicitly refer to this practice but his report in the appendix must have come from his own log just as Nansen's book does.

A very different type of RSP involved firing a cannon on special occasions such as when the two men leave the Fram to try and reach the North Pole (p. 370). While no reference to the firing of cannons is made during the two-man journey, Sverdrup also refers to this RSP when Nansen and Johansen leave and on other occasions involving May 17, Norwegian Independence Day, and the day the Fram finally reaches the open sea towards the end of the expedition (pp. 603, 614, 615, 644, 655). In a different vein, many people enjoyed the company of the dogs and puppies. In the first part of the journey Nansen refers various times to walking and playing with the dogs (pp. 138, 236, 248, 265–266). Not surprisingly he does not do so when attempting to reach the Pole, but Sverdrup does tell us that during excursions on the ice puppies would follow the crew and after rough weather they and the dogs would leave the ship to get some exercise (pp. 628, 639).

Another more personal practice involved smoking. While not mentioned on the two-man mission or by Sverdrup Nansen refers a number of times to the crew smoking cigars, cigarettes, or pipes both during normal social gatherings and more special events on the Fram (pp. 115, 134, 179–182, 184, 193, 204, 206, 211, 255, 260, 264, 281–283, 289, 290, 313–316, 319, 320, 324, 332, 333). In a different vein, Nansen reveals how important special foods, treats, and drinks were to the crew (Sverdrup does not discuss this topic). On the Fram the men enjoyed treats such as Turkish coffee, figs, raisins, almonds, mixed fruits, and honey cakes on different occasions ranging from banquets, Christmas, and other festivities, to waiting on the ice to see if the Fram would be crushed by the extremely packed ice (pp. 189, 199, 211, 281, 282, 287–290, 332, 333). On the two-man mission Nansen and Johansen celebrated various events such as reaching open water or shooting a bear by eating some chocolate (pp. 456, 457, 462, 463, 464, 476, 531). And after sledging all day, they would follow up supper with a "little extra drink" of hot water with whey powder that was "wonderfully comforting (p. 376)."

A different RSP involved photography. To begin with, a number of photos are included in Nansen's memoir especially when he is on the Fram. While the photographers are not identified, a reasonable guess is that Nansen took at least some of the pictures. During the second part of the expedition references are made to Nansen taking photos of gulls on a cliff, Johansen and a walrus, their winter hut, and walruses (pp. 486, 514, 551, 557). And when they are kayaking in the latter part of the mission and a walrus attacks them, Nansen fears that "water has got into the photographic apparatus, and perhaps my precious photographs are ruined (p. 569)." Lastly, Sverdrup comments once on how he and the doctor went out "photographing" a long way from the ship (p. 629).

Another significant practice, at least for part of the crew, involved reading books obtained from what Nansen refers to as the "library (p. 534)." In fact, a photo is included of the "reading-room" on the Fram with Nansen and three other crewmembers (p. 129). All the comments about this RSP are provided by Nansen and mostly refer to his reading different types of works although several remarks are about the crew reading books and a newspaper for the crew prepared one time by the doctor (pp. 134, 144, 148, 160, 164–166, 179–183, 185–190, 193, 195, 197, 203, 211, 260, 264, 265, 271, 283, 313, 324). Moreover, Nansen says, they longed for a book on the two-man trip since all they had to read were navigation-tables and an almanac (pp. 455, 456, 534). Yet he emphasizes that they read these two volumes many times and it was "still a comfort to see these books (p. 534)."

Music was also quite important to the men on the Fram as Nansen and Sverdrup call attention to. While Nansen mentions his or others playing the organ or simply making music, the majority of the comments refer to music making by other crewmembers on particular occasions such as rejoicing about how far they have traveled, birthdays, Christmas, New Years, or national holidays (pp. 115, 132–135, 154, 163, 179–182, 186–189, 227, 228, 250–252, 287, 288, 313–316, 319, 320, 324). Oftentimes their organ provided the music, but other kinds of musical activities were also enjoyed by the group such as playing violin or accordion, a violin recital, concerts, people singing, and even dancing to the songs being played. Sverdrup also emphasizes, how organ music and dancing enlivened their May 17 festival (pp. 614, 615). Lastly, the breadth of the music they enjoyed is very apparent when Nansen discusses how the organ was repaired when it stopped working, which allowed them to "go ahead with music sacred and profane, especially waltzes (p. 228)." Clearly the playing of music was a moving and energizing RSP for the men.

At the same time the crew participated in various outdoor and indoor recreational activities. Nansen describes ritualized outdoor activities on the Fram such as dog sledging, revolver practice, rifle practice, exercising, the doctor playing ball, a shooting competition, several crewmen sailing a small boat in a pool of water around the ship, and ice blasting to loosen up some extremely packed-up ice around the ship (pp. 153, 178, 186, 209, 211, 213, 236, 237, 239, 254, 264, 278). And while no reference is made to any kind of recreational practices on the two-man trip Sverdrup ordered everybody to exercise outside every day when conditions permitted it (pp. 606, 613, 634, 639). Recreational activities also occurred on a regular basis inside the Fram. Nansen tells us the crew played darts, games such as chest, dominoes, and Halma (a board game resembling Chinese checkers), different card games including Whist on many evenings, and enjoyed the newspaper the doctor prepared (pp. 156, 164–166, 168, 183, 184, 186, 192, 206, 211, 264, 287, 288, 327). Sverdrup also describes a type of craze that spread among the crew for a period of time in which the men shaved their heads in addition to their often betting on occurrences such as the drift of the ice (pp. 613, 614, 640). Both of these were recreational activities that provided some entertainment for the group.

Nansen also describes an RSP that he and most likely the other men engaged in: the observing of and appreciation for nature. Nearly all of his commentary about this ritualized activity on the Fram focuses on his own personal behavior. At times he refers to his appreciating the colors of his environment, the sky, and the snow, but the large majority of his remarks concerns the striking beauty of the aurora borealis or northern lights which he is fascinated by (pp. 160, 162–164, 167, 187–189, 222, 224, 228, 264, 280, 285, 286, 288, 289, 293, 325, 341). While Sverdrup doesn't address this topic, Nansen also refers to his looking at the shooting stars, open seawater, and the northern lights on his trip with Johansen (pp. 525–529, 531).

While all the previous RSPs were to varying degrees important to the crew, the last three we will examine are referred to in very large numbers and often in a way that indicates how crucial they were for everyone. The hunting of animals is one such practice that is discussed in all the phases of the expedition. While animals such as reindeers are pursued most commentary emphasizes how individuals, groups of two or three men, and even larger groups hunted bears, different birds, seals, and walruses. As previously mentioned, hunting clearly had utilitarian value since it provided extra food for the group, and in a limited number of cases protected them if they were attacked. But beyond that hunting was a ritualized practice that was symbolically meaningful and emotionally gratifying to crewmembers. It provided a break from their day-to-day lives, allowed them to compete with each other while enjoying a sense of adventure, and was fervently engaged in by many of the crew.

Nansen's account of hunting, while on the Fram supports these points (pp. 69, 70, 71, 93, 94, 104, 105, 107, 112, 116, 117, 135, 136, 144–147, 152, 153, 177, 178, 223, 224, 230, 256, 261, 266, 291, 292). For instance, he stresses how in hunting one experiences a "passion of the chase" and it is a "sport" that is enjoyed by the group (pp. 93, 135). On the North Pole attempt, he also refers numerous times to the two men hunting (pp. 429, 432, 448, 450, 452, 453, 456, 457, 462, 472, 473, 482, 492, 497, 501–508, 514, 516–519, 522, 540, 544, 547). And he declares: "A wonderful thing this love of sport; it is like setting fire to a fuse (p. 462)." Sverdrup also discusses crewmembers hunting and its benefits for them with some pages containing multiple comments (pp. 609, 617, 623, 640, 641, 648–651). He approvingly states that "this hunting life" has "a beneficial effect upon our spirits" and "appetite" and it provides us "encouragement and good living (p. 651)." Their enthusiasm for this RSP is clearly evident in their killing seventeen bears and probably hundreds of birds during their last summer of the expedition (pp. 648–650).

As popular as hunting was the next RSP, walking and snowshoeing, was as much if not more popular in terms of how often it is mentioned. While more references are made to walking, a significant proportion of commentary is devoted to snowshoeing, which was more suitable for certain conditions. Nansen engaged in the two by himself much of the time but also with others (pp. 84, 85, 138, 146, 154, 168, 174, 178, 179, 182, 184, 196, 202, 203, 205, 206, 209, 211, 217, 218, 220–222, 224, 236, 237, 258, 259, 264–266, 271, 277, 278, 281–284, 287–289, 294, 295, 312, 335, 340). He and other crewmen went on both short and long excursions. In fact, he states that he took "a stroll every morning" in addition to longer trips (p. 340). Nansen also made everyone go snowshoeing two hours daily and believed that it was a "good exercise," "great pleasure," and "everyone seemed to thrive on it (pp. 277)." Even on the two-man trip he participated in the same RSP (pp. 455, 456, 522–530, 558). For example, while in their hut the Artic winter often made walking extremely difficult and uncomfortable, yet Nansen managed to do so as much as possible. What is more, Sverdrup felt the same way about walking and snowshoeing (pp. 606, 607, 623, 628, 629, 634, 636, 637, 639). He too required daily snowshoeing and says he often walked four or five hours a day on the ice and felt, it "freshened up both soul and body (p. 634)."

In sum, all evidence indicates that walking and snowshoeing were pleasurable, invigorating diversions from normal ship life. They were excellent exercises that enabled the men to see the remarkable, changing Arctic environment including the sky and Northern Lights. And depending on the situation they allowed individuals to be both alone or in the company of others.

The last RSP normally enacted by all the crewmembers is special collective events involving celebrations. In the first part of the expedition, Nansen references a number of such events that occurred on a regular basis on the Fram (pp. 115, 132–135, 151, 153, 179–183, 187–189, 206, 207, 211, 213, 250–252, 281, 282, 287, 288, 289, 290, 313–316, 319, 320, 324, 333, 345, 350, 351, 353, 354). These collective gatherings celebrated occasions such as arriving at the New Siberian islands in the first part of the journey, reaching new northerly latitudes, anniversaries for the age of the Fram (e.g., one or two years old), different individuals' birthdays, the first appearance of the sun after winter, the day before Christmas, Christmas day, New Years Eve, several farewell parties when Nansen and Johansen leave on their trip, and May 17, Norway Memorial Day (anniversary for the Norwegian Constitution).

On these occasions the crew engaged in different practices. While the number and nature of these actions varied among events, crewmembers often participated in many if not all of them. These activities included magnificent dinners, delicious desserts, fruits, sweets, and nuts, drinks (e.g., beer, ale, liquors, punch), decorating the main meeting room on the Fram, toasts, speeches, reading old newspapers aloud, giving thanks to each other, praising the Fram and drinking to her, smoking cigars, cigarettes, and pipes, humorous prizes, gifts, music usually involving the organ or violin, concerts, singing, dancing, poems, verses, darts, cards, merriment, laughter, and parading often dressed in humorous outfits and/or accessories on the ice and around the ship. These celebrations could last for long periods of time. Some began during the day or in the evening and went late into the night.

Of particular note are the foods eaten during these special events and for that matter on a daily basis. Nansen's foresight and extremely careful planning for the expedition included food supplies that were of an extremely high quality. This meant the cooks could prepare very satisfying, tasty daily meals and exceptional dinners for the celebrations. To be more specific servings for celebrations included amongst other things some of the following foods: ox-tail soup, fish pudding, minced fish with curried lobster, potatoes and melted butter, sausages, tongue, roast of deer, pork cutlets, turtle, fried halibut, boiled mackerel, French beans, pease, cranberry jam, cloudberries with cream, mango chutney, tinned pears and peaches, dried bananas, figs, raisins, Norwegian wild strawberries, ryebread, and gingerbread.

As for regular meals, Sverdrup gives us a good idea about at least some of the food supplies that were undoubtedly used everyday when he lists the provisions that were stored for an emergency, which would have required a lengthy sledging trip or an extended stay on the ice. Examples of such food items are beef, corned beef, ham, mutton, collops (slices of bacon), vegetables, groats (grains such as oat, wheat, barley, and rye), danish butter, preserved fish, jellies, pemmican, potatoes, breads, chocolate, milk, flour, sugar, and so on (pp. 624, 625).

Sverdrup also refers to special celebratory events on the Fram (pp. 613–615, 632, 633, 635, 639, 641, 642, 644). He tells us how the crew celebrated Whitsun eve (the seventh Sunday after Easter), May 17, Norway Memorial Days, how long they had travelled on the Fram, the construction of an ice hut observatory, reaching new westerly longitudes, the return of the sun, crewmen's birthdays, a one-year anniversary for the beginning of the two-man journey, and occasions when Sverdrup's popular claret (red wine) was served. Here too the men participated in a variety of activities many of which are similar to those discussed by Nansen. These practices included enjoying different delicacies for supper, a house-warming for the new observatory hut followed by humorous speeches that were met with great enthusiasm and high spirits, decorating the main meeting room with flags for a little feast, and a better than usual dinner, coffee afterwards, and a punch bowl in the evening. And on one of the Norway Memorial celebrations activities included parading with flags on the ice with accordion music, cheering, firing the ship's guns, a splendid dinner with alcoholic drinks, and an outdoor afternoon festival with gymnastics, shooting at hares, and so on followed by an excellent supper and then a bowl of punch.

Lastly, we learn that despite the severe and challenging conditions Nansen and Johansen faced, they managed to engage in smaller versions of some of the RSPs found on the Fram (pp. 388, 390, 392, 408, 409, 412, 413, 455, 456, 462, 468, 476, 484, 530–531). Special events celebrated involved reaching their northernmost location, Easter, Johansen's birthday, May 17, Midsummer Day, how long the expedition had taken, and they had been on their two-man journey, a large rain melting the snow, sighting land, reaching open water, and both Christmas Eve and Christmas day. On these occasions, the two men engaged in a number of practices such as a feast made up of lobscouse (a stew comprised of meat, vegetables, and hardtack), bread and butter, chocolate, and a hot whey drink, a special dinner made up of hot whey and water, fish au gratin, whortleberries, and lime juice grog, another dinner of lobscouse and lime-juice grog, flying flags on their sledges and having a first-rate dinner of lobscouse, whortleberries, and a juice drink in honor of May 17, and on another occasion participating in festivities such as studying their charts, reading, walking, and relishing a delightful meal of pancakes with sugar.

Further activities included having a piece of chocolate, greatly enjoying a cup of chocolate with raw blubber, partaking of a feast of lobscouse made of potatoes, pemmican, dried bears' and seals' flesh and bear tongues chopped up together, followed by bread-crumbs fried in bears' grease, a bit of chocolate, etc., and flying the Norwegian flag and eating a great tasting lobscouse made of pemmican and potatoes. Lastly, the two had a festive Christmas Eve eating for supper fiskegratin made of powdered fish, maize meal, and train-oil (oil obtained from the blubber of a whale) along with fried bread for dessert followed by chocolate and bread the next day.

All of these collective celebrations involving the foods they prepared along with the flags they flew were very emotionally satisfying to the two men and enhanced their feelings of pride. At the same time, they created a sense of connection with their homeland and comrades on the Fram, reinforced the bond between Nansen and Johansen, and lifted their morale. Of course, much the same could be said for all the celebratory events on the Fram.

To conclude, we have identified 17 RSPs engaged in, within the three parts of the expedition. Certain rituals are referred to a relatively small number of times such as telling stories and yarns, shooting the guns on the ship (although this was probably a very salient activity), and playing with and enjoying the company of the dogs and pups. Most other RSPs ranged from moderately high levels of references, e.g., reading, music, and indoor and outdoor recreational activities, to extremely high levels, e.g., work rituals, observing nature especially when Nansen is on the Fram, hunting, walking and snowshoeing, and special collective events. In essence, all the evidence indicates that, notwithstanding the fact that extreme conditions restricted what kind of ritual practices could be enacted on the two-man trip, crewmembers as a whole engaged in many RSPs that were very salient, often participated in, well supported by various sorts of resources, and similar in that they usually emphasized the value of work focused on the needs of the expedition, the ship, and crew along with scientific research, maintaining positive relations among everyone, and/or upholding the physical health and high morale of every crewmember. In terms of the conceptual framework guiding this investigation most RSPs exhibited a high, if not extremely high rank which means these practices had a significant impact on individuals and their interaction with each other.

We also find that crewmembers engaged in a diverse array of RSPs ranging from taking photographs to reading, smoking, hunting, card playing, eating treats, and snowshoeing. Most were reconstituted ritual practices, but some were also adjusted to their radically new environment whether that involved observing the aurora borealis throughout the day or parading on the ice and around the ship on certain occasions. Furthermore, RSPs ranged from individualized/personal activities such as diary writing to informal social interaction on both the ship or ice and more formally organized undertakings, a prime example of which would be the larger celebrations which could go on for hours and involve exceptional dinners, concerts, speeches, festive activities and games, and so on.

These celebrations were also special collective events that produced strong, positive emotions among crewmembers. Numerous descriptions of these events suggest that individuals' attention was focused on particular persons, actions, objects, and symbols (e.g., flags, decorations in the main meeting room, and recitals). The rate of interaction also appears to have occurred at a moderate if not high level as they often engaged in an animated fashion in activities such as singing, dancing, and gathering later in the evening around the punch bowl. And interdependence existed between all, since everyone was involved in the event, oftentimes doing different things, each of which contributed to the overall endeavor, e.g., playing musical instruments, singing, giving speeches, making toasts, preparing the meals, dancing, reciting verses, providing comedic entertainment. Moreover, resources were clearly available, due in large part to Nansen's planning, such as foods, materials for costumes, musical instruments like the organ, liquors, flags, necessary items for playing games or holding shooting contests, and so on. So too there was a definite sense of co-presence in which people were aware of each other's participation in the event whether they were in the meeting room of the ship, on the ice, engaging in an activity that directed their attention to each other, or sitting together in a small hut. Taken together these factors created a significant amount of positive collective emotions, which in turn enhanced actors' feelings of solidarity with others and commitment to the group.

Lastly, some of these emotionally charged events helped to engender a sense of collective pride. Celebrations for how far the Fram had travelled, how long they had been at sea and in the ice, how successfully the Fram was performing, and their homeland, Norway, directed attention back to the crew enhancing a pride in themselves and their accomplishments. This most likely strengthened group cohesion and boosted the crew's morale.

To sum up, all of these RSPs produced a highly effective program of ritualized practices and collective events. Notwithstanding the very real frustrations and disappointment of Nansen and the others when the ship was drifting slower than desired this assortment of activities shaped in many ways the lives of all crewmembers. Moreover, Nansen and Sverdrup played a major role in creating and maintaining this program through the entirety of their Arctic venture. In a rigorous and thorough manner, Nansen prepared for all facets of the mission including the two-man journey. By all accounts this was the approach he took in addressing the needs of the crew.

Finally, we find that a number of factors influenced the performance of RSPs and the success of the mission. As in certain of the previously examined ventures this expedition was highly organized and carefully planned. The selection of the crew by Nansen was also conducted in a thoughtful manner. Evidence indicates that crewmembers were highly motivated workers, convivial in nature, and dedicated to the undertaking. In this regard, Nansen's faith in the mission and its success was a source of inspiration to many, if not all of the crew. Furthermore, the group was homogeneous in that all were from Norway except for one individual who was Swedish. Their shared cultural background gave them a familiarity with many of the rituals they took part in and their leaders who were Norwegian. And many of these RSPs were engaged in from the very beginning of the mission, which laid the foundation for subsequent practices throughout their journey.

We should also appreciate that a well-defined authority structure existed involving Nansen and Sverdrup who were respected and liked by the crew. Both individuals were experienced polar explorers who successfully communicated with and cooperated with each other. It is also noteworthy that while Nansen was unequivocally recognized as the ultimate authority his style was one in which he eagerly shared the burdens, constraints, and joys of the crew whether that involved, for instance, his hunting and exercising like the others, eating meals with them, sharing in their labors, and being an active participant in all special collective events and celebrations, both on the Fram and the trek with Johansen.

The result was a highly successful expedition, which in the face of major challenges, persevered and left a lasting mark on the world of polar exploration.

Conclusion

These four missions demonstrate in a powerful way how ritual practices contribute to the well-being, morale, and commitment of crews on polar expeditions. In the final chapter, we summarize this study's findings and discuss some of their implications for different types of hazardous ventures.

6
CONCLUSION

Drawing upon SRT, we have examined the role ritual plays in 19 polar expeditions over a 100-year period. Our focus is squarely on the personal and social dimensions of these hazardous ventures.

In doing so, the purpose of this investigation is not to blame or show disrespect for any member of these expeditions, e.g., leaders, officers, or crewmembers. The goal is to analyze an important dimension – ritual dynamics – of these missions by examining issues such as whether crewmembers cope with the situation they are in, are rituals promoted in expeditions, and the impact they have on individuals.

This study does not simply examine one or two cases. When only a few expeditions are examined, it can be difficult to recognize the value of rituals and how their presence or absence may influence crews. But when multiple cases are examined the importance of RSPS becomes much more apparent. Patterns can be identified across multiple cases involving ritual dynamics and outcomes for crews in terms of social conditions and morale.

At the same time there are other sources we might have examined but didn't. For instance, other examples of successful expeditions could include the Amundsen ([1912] 1976), M'Clintock (1860), and Simpson (1957) expeditions. There is a limit, however, on how many accounts could be included in this study both for practical reasons such as the time and effort it takes to conduct such an investigation and because the cases examined are more than sufficient for carrying out the analysis.

It is important to emphasize that in examining the RSPs occurring in an expedition, how these practices are or are not facilitated, and their consequences for the members of the mission, we often find that there are several

DOI: 10.4324/b23044-6

factors, which influence the social dynamics of crews. This can make it difficult, if not impossible, to separate all the elements in a clear-cut manner or simple temporal fashion. They can all occur simultaneously. For instance, the hazardous conditions an expedition faces, ineffective leadership, social prejudice, and inadequate preparation could all contribute at the same time to the absence of solidarity enhancing RSPs and divisiveness within a group.

But the key point is that however it plays out, the presence or absence of RSPs is a crucial part of the social dynamic that contributes to a mission's success or failure in a social psychological sense. They are a key element in the social equation that defines what life is like for crewmembers.

This is why a qualitative analysis of accounts can be extremely useful. It lets us examine in an in-depth manner the complexities and nuances of the interpersonal relations of crews, what is meaningful to them, and how rituals can affect their thoughts, feelings, and behaviors. Indeed, the evidence collected in this study shows the multiplicity of ways RSPs may be engaged in, and are beneficial to crewmembers. This point is further buttressed by the observations of different leaders and authors of accounts, which substantiate the crucial role such practices play in contributing to the high morale, dedication, and social equanimity of crews.

With that said our analysis of these accounts has led to the grouping of the 19 expeditions into three types: failed, successful, and extremely successful missions. Several points need to be made about these findings.

To begin with, it is important who participates in RSPs. For instance, in failed expeditions sometimes individuals and/or subgroups do take part in RSPs, including the author of the account who may engage in a collection of quite consequential rituals, i.e., a personal program of RSPs. Yet most of the crew participate in few if any RSPs. While important for those individuals who engage in rituals, these are still considered failed expeditions since the focus is on the crew as a whole.

As for the rituals themselves, SRT argues that salience, repetitiveness, homologousness, and resources are the factors, which influence the rank or dominance of RSPs. This of course means that rituals can vary in their significance and impact.

Rituals can also greatly vary in terms of content or type. As we have seen, various crews engage in numerous and sometimes very different RSPs. They include work activities such as running and maintaining a ship or conducting scientific tasks, religious practices, sporting events, parties, reading, diary writing, hunting, recreational and playful activities such as games, playing cards, singing and dancing, and listening to or talking on radios, celebrating birthdays and important occasions, special dinners, watching movies, and so on. Additional RSPs involve taking walks, exercising, storytelling, conversation and debate (including discussions about food), personal informal work projects, teaching or taking classes, staging shows for entertainment, creating newspapers, painting pictures and taking photographs, sewing, smoking, and periodically enjoying special treats such as hot milk or candy.

What is more, we find that there is not necessarily a set of specific kinds of RSPs that are required for coping with the challenges of an expedition other than basic work practices which help achieve the goals of exploration, physical survival, and oftentimes research. Rather various rituals can be beneficial to crews and help them deal with their situation. The important point is that dominant, meaningful, and consequential rituals have to be engaged in whatever their specific content. For example, in one successful expedition religious practices or hunting may be of central importance while in another we find little or no references to such activities but rather repeated commentary about parties, shows, and special dinners.

Rituals also differ in terms of whether they involve personal activities, informal group/social practices, or formal/quasi-formal organized activities. What is more, RSPs can be reconstituted from people's pre-expedition lives or be of a more novel nature whether that involves creative modifications of older rituals or completely new practices.

Moreover, expeditions differ in terms of whether collective ritual events occur and the degree to which they exhibit varying degrees of emotional intensity, i.e., collective emotions. Various accounts show that collective events, which generated positive emotions such as satisfaction, happiness, and joy occurred in certain missions. And it appears that the four factors discussed in Chapter 2 influenced the collective emotions some crews experienced. While evidence does not allow for a precise analysis of the role played by those factors, a number of the accounts are highly suggestive. For instance, one type of event with a very high emotional intensity was special celebrations which usually took hours and involved activities such as plays, music, singing, gift-giving, toasts, speeches, grand dinners, drinking, games, contests of various sorts, and so on. They were often characterized by the crew's high focus of attention on these activities, high interdependence due to the numerous activities needed to stage the events and the participation of most, if not all crewmembers, ample resources ranging from food and musical instruments to a high degree of co-presence because of the limited physical space in which many events occurred (e.g., in their ship), and intense, frequent interaction.

As for collective pride there are fewer cases compared to collective emotions but on certain missions there clearly were reflexive RSPs that directed the crew's attention to themselves and boosted their sense of worth or pride as in the North Pole trek led by Peary (described by Henson) and the Nansen expedition.

As a very broad generalization when we move from failed to successful and extremely successful expeditions, we find more powerful, highly ranked rituals, increases in the types of practices participated in, more personal/informal group/organized rituals, more reconstituted and in some cases new rituals, increasing numbers of collective events marked by strong collective emotions, more instances of collective pride among the crews, and carefully developed programs of RSPs. But there are very clear variations on all of this. Crews are not the same and differences do exist in terms of the RSPs engaged in. In other words, ritual dynamics can play out in quite different ways even among failed, successful, or extremely successful missions.

At the same time, there are different factors that may or may not facilitate significant RSPs and ultimately high morale, positive social conditions, and productive group dynamics. Such factors include how well or poorly expeditions are organized, quality of preparations for the mission, whether there is a clear structure of authority, the degree to which authority figures are unified, the quality of the crew selection process, how clearly authority figures communicate with the crew, composition of crews (e.g., nationality, race, ethnicity, gender, personnel differences such as military, scientists, sailors), homogeneity of crew or if not homogeneous whether efforts are made to unite different subgroups, crewmembers' and leaders' experience on previous missions, their training and motivation, previous relations between individuals, i.e., worked together on prior expeditions, social biases/prejudices or lack thereof, temperament of individuals, and whether opportunities exist to interact, communicate, and share RSPs with others who are not part of the expedition.

Further factors include how effective leaders are, whether RSPs are of high rank in terms of salience, repetitiveness, homologousness, and resources, and are strengthened through collective events, collective emotions, and/or collective pride, if RSPs are encouraged in the beginning of the mission, thus, laying a foundation for subsequent RPSs, and whether programs of RSPs are fostered by leaders and possibly others. For instance, very successful leaders usually promoted strong programs of ritual practices and collective events as in the case of Ronne, Scott, Shackleton and Wild, and Nansen.

This study of polar expeditions also has a number of implications for our understanding of ritual dynamics in different kinds of situations. By focusing on this dimension of social behavior our research directly contributes to SRT, especially work dealing with the importance of disruption, deritualization, and reritualization and how the latter helps people cope with breakdowns in rituals that help shape our normal lives. As such this investigation contributes to a growing body of research and concepts addressing these concerns as discussed in Chapter 2.

Furthermore, several directions for future research are possible. For instance, a similar approach could be used to investigate the role of ritual in polar expeditions from 1950 up to the present. These undertakings involve both exploratory missions and research bases which have substantially increased in number and size since the mid-twentieth century. Moreover, some of these bases are, or will likely soon be, of a commercial nature focused on oil drilling and production or of a military nature, which may impact how rituals are enacted and organized. Another possibility would involve the comparative study of ritual dynamics in polar expeditions and very different settings such as concentration camps and POW camps where disruption and deritualization are due to coercion (Knottnerus 2002; Van de Poel-Knottnerus and Knottnerus [2011] 2016).

Another topic deserving further examination involves the role of leadership in strategically utilizing RSPs. Some of the lessons from our investigation of polar missions might be used to develop a schema for helping leaders create successful ritual programs in different settings. Such an analysis could be relevant for not only leaders of expeditions and other hazardous ventures but also formal organizations and more informal groups.

The study of expeditions can also inform research focused on a very different phenomenon: disasters. Calamities due to nature (e.g., tornadoes, floods, earthquakes, hurricanes) or humans (e.g., terrorism, nuclear reactor malfunctions, chemical plant accidents) are another cause of disruption and deritualization. Here too, the ritual dynamics of those affected by these kinds of events are critically important involving issues such as whether people reconstitute RSPs after disasters, what kind of rituals they might engage in, and how beneficial such actions are for communities and individuals (Johnson, Knottnerus, and Gill n.d.).

In a different vein, findings from this study could inform the analysis of social change, its impact on people, and social persecution. When social change occurs involving, for instance, economic loss, dislocations within communities, and the breakdown of social arrangements and/or everyday RSPs this may create uncertainty, stress, anxiety, a sense of threat, fear, anger, and other outcomes. In other words, social change can create disruption and deritualization in people's lives. This situation may then lead to those who are affected by such developments to identify, correctly or incorrectly, and attack attributed causes of the predicament they are suffering from. Simply stated, it can lead to the persecution of groups who are purportedly responsible for these changes. And the persecution of these groups may be promoted and exploited by political leaders who employ collective events and other ritualized activities. So too, collective events such as mass rallies, concerts, or celebrations can provide a sense of security, collective affirmation, and boost in morale for the group negatively impacted by social change. In doing so these events and shared emotions may reinforce the group's vilification and persecution of others. Issues such as these are being addressed in other research (Knottnerus n.d.).

Lastly, the findings and ideas presented here could be applied to other hazardous ventures and settings. Such groups might include military/combat units, submarine crews, isolated groups subject to high stress such as mountain climbing teams, and prisoners incarcerated in penitentiaries, including those placed in solitary confinement. Another setting involves refugee camps where people are separated from others and usually housed in very restrictive conditions for extended periods of time. Here too, the ability to engage in reritualization in order to cope with the disruptive experience of placement in such camps should be of concern not just to those living in them but also to those responsible for the management and well-being of these persons. The fact that tens of millions of people throughout the world reside in such camps and suffer from numerous social problems and mental disorders only adds to the importance of this issue (Chen 2016).

More generally these ideas are directly relevant to the argument that many kinds of situations and events involving disruption and deritualization can be the source of trauma, which can be ameliorated and healed through reritualization (Gordon-Lennox 2022).

We should as well remember that, while this study has focused on a 100-year span in the nineteenth and twentieth centuries, expeditions are not a thing of the past. In addition to polar missions, which continue to flourish, a new wave of hazardous ventures has begun. These endeavors involve the exploration of outer space along with probable commercial and military missions. While the international space station has been a prominent example of such missions it is almost a certainty that other ventures are quickly coming involving space travel, exploration, space habitation, and possibly space colonies. These endeavors involving extreme degrees of isolation, long periods of time, the restrictive nature of their environment, exposure to extremely dangerous conditions, and significant degrees of stress will encounter many of the same challenges faced by polar expeditions.

These challenges are already being faced by those on the international space station and simulations here on earth involving groups of people living together for extended periods of time, as will be the case on planets such as Mars. The long journey to and from Mars and the astronauts' stay while researching the planet are probably the most often discussed examples of future undertakings. But many other possibilities exist. Serious attention is already being given to operations ranging from establishing bases on the moon by countries such as China and the United States to the development of orbiting lunar outposts. Other possible projects include spaceports where large numbers of people live in space colonies. Furthermore, various countries are considering establishing colonies on Mars in the coming century some of which may be collaborative missions involving different nations.

These and other kinds of ventures involving both extended travel and habitation on planets, asteroids, or human made environs in space will face very serious challenges including the disruption of RSPs that influence people's lives and the need to reconstitute old and/or new ritual practices in their new non-earth surroundings. Other challenges will also confront these explorers such as the potential problems created by long-term collaborative ventures composed of people from different countries. For instance, there will be a great need, as in some polar expeditions, to develop RSPs that are meaningful to the members of these mixed groups, and which facilitate communication, cooperation, positive emotions, social cohesiveness, and high morale. Such practices should help bridge the gap between people from different cultures, thereby, serving to unite them in their isolated and inhospitable world.

To conclude, this study has relevance for various situations and areas of interest. That is why this book was written in a manner that would hopefully be of value to not only sociologists, social psychologists, and other social scientists but those with interests in polar exploration, other types of expeditions, leadership in hazardous ventures, Naval and military history, and various organizations such as prisons, refugee camps, and other settings including nursing homes where people may experience extreme degrees of disruption and deritualization upon their entry into them. Indeed, this research should contribute to our understanding of numerous topics ranging from disasters and space exploration to social change and persecution.

In closing, the contributions and sacrifices of all those who participated in the polar expeditions examined here deserve our recognition and respect. Based on our study of these missions the value of ritual in influencing the experiences and group dynamics of these crews should also be appreciated. We find that over a 100-year period, some were able to use and greatly benefit from such activities whether this was planned, based on intuition, grounded in folk knowledge, or for some other reason.

Hopefully, future endeavors will even more fully recognize the value of ritual whether that involves engaging in the same or different RSPs in similar or other kinds of hazardous ventures marked by isolation, stress, and the disruption of those practices that define our everyday lives.

BIBLIOGRAPHY

Albanov, Valerian. 2000. *In the Land of White Death: An Epic Story of Survival in the Siberian Arctic*. New York: The Modern Library (Random House).

Amundsen, Roald. 1976 [1912]. *The South Pole*. New York: St. Martins.

Armstrong, Alexander. 1857. *A Personal Narrative of the Discovery of the North-West Passage*. London: Hurst and Blackett.

Barr, Willliam. 2007. *Arctic Hell-Ship: The Voyage of HMS Enterprise 1850–1855*. Edmonton, Canada: The University of Alberta Press.

Bell, Catherine. 1992. *Ritual Theory, Ritual Practice*. Oxford: Oxford University Press.

Berger, Peter and Thomas Luckmann. 1966. *The Social Construction of Reality: A Treatise in the Sociology of Knowledge*. New York: Anchor Books.

Bhandari, Roshan Bhakta, Norio Okada and J. David Knottnerus. 2011. "Urban Ritual Events and Coping with Disaster Risk: A Case Study of Lalitpur, Nepal." *Journal of Applied Social Science* 5:13–32.

Blake, E. Vale. 1874. *Arctic Experiences: Containing Capt. George E. Tyson's Wonderful Drift on the Ice Floe, A History of the Polaris Expedition, the Cruise of the Tigress, and Rescue of the Polaris Survivors*. New York: Harper & Brothers.

Bocock, Robert J. 1970. "Ritual: Civic and Religious." *The British Journal of Sociology*, 21(3):285–297.

Bryant, Joseph M. 1994. "Evidence and Explanation in History and Sociology: Critical Reflections on Goldthorpe's Critique of Historical Sociology." *The British Journal of Sociology* 45:3–19.

_____. 2000. "On Sources and Narratives in Historical Social Science: A Realist Critique of Positivist and Postmodernist Epistemologies." *British Journal of Sociology* 51:489–523.

Chen, Sophia. 2016. "Quiet Epidemic of Mental Disorders in Refugees." http://www.realclearscience.com/2016/02/17

Cherry-Garrard, Apsley. 1965. *The Worst Journey in the World: Antarctic 1910–1913*. London: Chatto and Windus.

Collins, Randall. 2004. *Interaction Ritual Chains*. Princeton, NJ: Princeton University Press.

Corner, George W. 1972. *Doctor Kane of the Arctic Seas.* Philadelphia: Temple University Press.

Decleir, Hugo (ed.). 1999. *Roald Amundsen's Belgica Diary: The First Scientific Expedition to the Antarctic.* Huntingdon and Norfolk: Bluntisham Books and Erskine Press.

Delano, Daisha Lee and J. David Knottnerus. 2018. "The Khmer Rouge, Ritual and Control." *Asian Journal of Social Science* 46:79–110.

Dion, Kenneth L. 2004. "Interpersonal and Group Processes in Long-Term Spaceflight Crews: Perspectives from Social and Organizational Psychology." *Aviation Space, and Environmental Medicine* 75 (Supplement 1): C36–C43(8).

Douglas, Mary. 1970. *Natural Symbols.* New York: Vintage.

Durkheim, Emile. 1965 [1915]. *The Elementary Forms of the Religious Life.* New York: Free Press.

Edwards, Jennifer and J. David Knottnerus. 2007. "The Orange Order: Strategic Ritualization and Its Organizational Antecedents." *International Journal of Contemporary Sociology* 44:179–199.

_____. 2010. "The Orange Order: Parades, Other Rituals, and Their Outcomes." *Sociological Focus* 43:1–23.

_____. 2011. "Exchange, Conflict and Coercion: The Ritual Dynamics of the Notting Hill Carnival Past and Present." *Ethnic Studies Review* 34:107–133.

Ellsberg, Edward. 1938. *Hell on Ice: The Saga of the "Jeannette".* New York: Dodd, Mead and Co.

Emerson, Richard M. 1966. "Mount Everest: A Case Study of Communication Feedback and Sustained Group Goal-Striving." *Sociometry* 29:213–227.

Giddens, Anthony. 1984. *The Constitution of Society: Outline of the Theory of Structuration.* Cambridge: Polity.

Glassner, Barry and Jay Corzine. 1982. "Library Research as Fieldwork: A Strategy for Qualitative Content Analysis." *Sociology and Social Research* 66:305–319.

Goffman, Erving. 1967. *Interaction Ritual: Essays on Face to Face Behavior.* Garden City, NY: Anchor Books.

Goldthorpe, John H. 1991. "The Uses of History in Sociology: Reflections on Some Recent Tendencies." *The British Journal of Sociology* 42:211–230.

Gordon-Lennox, Jeltje. 2022. *Coping Rituals in Fearful Times: An Unexplored Resource for Healing Trauma.* Switzerland: Springer Nature Switzerland.

Guan, Jian and J. David Knottnerus. 1999. "A Structural Ritualization Analysis of the Process of Acculturation and Marginalization of Chinese Americans." *Humboldt Journal of Social Relations* 25:43–95.

_____. 2006. "Chinatown under Siege: Community Protest and Structural Ritualization Theory." *Humboldt Journal of Social Relations* 30:5–52.

Haig-Thomas, David. 1939. *Tracks on the Snow.* New York: Oxford University Press.

Henson, Matthew A. 1989 [1912]. *A Black Explorer at the North Pole.* Lincoln: University of Nebraska Press.

Hocart, A. M. 1968 [1950]. *Caste: A Comparative Study.* New York: Russell & Russell.

Inglefield, Edward Augustus, 1853. *A Summer Search for Sir John Franklin with a Peep into the Polar Basin.* London: Thomas Harrison.

Jarvis, Tim. 2013. *Chasing Shackleton.* New York: William Morrow/Harper Collins Publishers.

Johnson, Kevin, J. David Knottnerus and Duane Gill. n.d. "Disasters and Rituals: An Analysis of Two Communities Impacted by Tornadoes."

Johnson, Phyllis J. and Peter Suedfeld. 2012. "Coping with Stress through the Microcosms of Home and Family among Arctic Whalers and Explorers." *The History of the Family* 1:41–62 (published online: 03 Jan 2012).

Kahn, Pauline Maki and Gloria R. Leon. 1994. "Group Climate and Individual Functioning in an All-Women Antarctic Expedition Team." *Environment and Behavior* 26:669–697.

Kane, Elisha Kent. 1857. *The United States Grinnell Expedition in Search of Sir John Franklin: A Personal Narrative*. Boston: Phillips, Sampson & Co.

Kertzer, David I. 1988. *Ritual, Politics, and Power*. New Haven, CT: Yale University Press.

Knottnerus, J. David. 1997. "The Theory of Structural Ritualization." pp. 257–279 in *Advances in Group Processes*, Volume 14, edited by Barry Markovsky, Michael J. Lovaglia and Lisa Troyer. Greenwich, CT: JAI Press.

_____. 1999. "Status Structures and Ritualized Relations in the Slave Plantation System." pp. 137–147 in *Plantation Society and Race Relations: The Origins of Inequality*, edited by Thomas J. Durant Jr. and J. David Knottnerus. Westport, CT: Praeger.

_____. 2002. "Agency, Structure and Deritualization: A Comparative Investigation of Extreme Disruptions of Social Order." pp. 85–106 in *Structure, Culture and History: Recent Issues in Social Theory*, edited by Sing C. Chew and J. David Knottnerus. Lanham, MD: Rowman & Littlefield.

_____. 2005. "The Need for Theory and the Value of Cooperation: Disruption and Deritualization." (Presidential Address, Mid-South Sociological Association, Baton Rouge, 2003). *Sociological Spectrum* 25:5–19.

_____. 2009. "Structural Ritualization Theory: Application and Change." pp. 70–84 in *Bureaucratic Culture and Escalating World Problems: Advancing the Sociological Imagination*, edited by J. David Knottnerus and Bernard Phillips. Boulder, CO: Paradigm Publishers.

_____. 2010. "Collective Events, Rituals, and Emotions." pp. 39–61 in *Advances in Group Processes*, Volume 27, edited by Shane R. Thye and Edward J. Lawler. Bingley: Emerald Group Publishing Limited.

_____. 2014a. "Emotions, Pride and the Dynamics of Collective Ritual Events." pp. 43–54 in *Understanding Collective Pride and Group Identity: New Directions in Emotion Theory, Research and Practice*, edited by Gavin Brent Sullivan. London and New York: Routledge.

_____. 2014b. "Religion, Ritual, and Collective Emotion." pp. 312–325 in *Collective Emotions: Perspectives from Psychology, Philosophy, and Sociology*, edited by Christian von Scheve and Mikko Salmela. Oxford: Oxford University Press.

_____. 2016 [2011]. *Ritual as a Missing Link: Sociology, Structural Ritualization Theory and Research*. New York and London: Routledge.

_____. n.d. *Preventing Persecution: Understanding Rituals and the Role of Groups*. New York and London: Routledge.

Knottnerus, J. David and Phyllis E. Berry. 2002. "Spartan Society: Structural Ritualization in an Ancient Social System." *Humboldt Journal of Social Relations* 27:1–42.

Knottnerus, J. David and David G. LoConto. 2003. "Strategic Ritualization and Ethnicity: A Typology and Analysis of Ritual Enactments in an Italian American Community." *Sociological Spectrum* 23:425–461.

Knottnerus, J. David, David L. Monk and Edward Jones. 1999. "The Slave Plantation System from a Total Institution Perspective." pp. 17–27 in *Plantation Society and Race Relations: The Origins of Inequality*, edited by Thomas J. Durant Jr. and J. David Knottnerus. Westport, CT: Praeger.

Knottnerus, J. David, Jason S. Ulsperger, Summer Cummins and Elaina Osteen. 2006. "Exposing Enron: Media Representations of Ritualized Deviance in Corporate Culture." *Crime, Media, Culture: An International Journal* 2:177–195.

Knottnerus, J. David and Frédérique Van de Poel-Knottnerus. 1999. *The Social Worlds of Male and Female Children in the Nineteenth Century French Educational System: Youth, Rituals and Elites.* Lewiston, NY: Edwin Mellen Press.

Knottnerus, J. David, Jean L. Van Delinder and Jennifer Edwards. 2016 [2011]. "Strategic Ritualization and Power: Nazi Germany, the Orange Order, and Native Americans." pp. 73–105 in *Ritual as a Missing Link: Sociology, Structural Ritualization Theory and Research,* edited by J. David Knottnerus. New York and London: Routledge.

Lansing, Alfred. 1959. *Endurance: Shackleton's Incredible Voyage.* New York: McGraw Hill Book Company.

Leon, Gloria R. 1991. "Individual and Group Process Characteristics of Polar Expedition Teams." *Environment and Behavior* 23:723–748.

Leon, Gloria R., Carl McNally and Yossef S. Ben-Borath. 1989. "Personality Characteristics, Mood, and Coping Patterns in a Successful North Pole Expedition Team." *Journal of Research in Personality* 23:162–179.

Liang, Bin, J. David Knottnerus and Michael A. Long. 2016. "A Theoretical Model of Drug/DUI Courts: An Application of Structural Ritualization Theory." *American Journal of Criminal Justice* 41:31–46.

Lin, Xiaohua, Jian Guan and J. David Knottnerus. 2011. "Organizational and Leadership Practice of Micro Ethnic Entrepreneurship in Multicultural Context: A Structural Reproduction Analysis." *International Journal of Business Anthropology* 2:48–65.

Lowenthal, Leo. 1961. *Literature, Popular Culture, and Society.* Englewood Cliffs, NJ: Prentice-Hall.

——. 1986. *Literature and the Image of Man, Communication in Society, Volume 2.* New Brunswick, NJ: Transaction Books.

Lustick, Ian S. 1996. "Historiography, and Political Science: Multiple Historical Records and the Problem of Selection Bias." *The American Political Science Review* 90:605–618.

M'Clintock, Sir Francis Leopold. 1860. *The Voyage of the 'Fox' in the Arctic Seas: A Narrative of the Discovery of the Fate of Sir John Franklin and His Companions.* Boston: Ticknor and Fields.

M'Clure, Robert. 1857. *The Discovery of the North-West Passage.* 2nd ed., edited by Osborn, Sherard, London: Longman, Brown, Green, Longmans, & Roberts.

McKinlay, Willam Laird. 1999 [1976]. *The Last Voyage of the Karluk: A Survivor's Memoir of Arctic Disaster.* New York: St. Martin's Griffin.

Meij, Jan-Martijn, Meghan D. Probstfield, Joseph M. Simpson and J. David Knottnerus. 2013. "Moving Past Violence and Vulgarity: Structural Ritualization and Constructed Meaning in the Heavy Metal Subculture." pp. 60–69 in *Music Sociology: Examining the Role of Music in Social Life,* edited by Sara Towe Horsfall, Jan-Martijn Meij and Meghan D. Probstfield. Boulder, CO: Paradigm Publishers.

Minton, Carol and J. David Knottnerus. 2008. "Ritualized Duties: The Social Construction of Gender Inequality in Malawi." *International Review of Modern Sociology* 34:181–210.

Mitra, Aditi and J. David Knottnerus. 2004. "Royal Women in Ancient India: The Ritualization of Inequality in a Patriarchal Social Order." *International Journal of Contemporary Sociology* 41:215–231.

_____. 2008. "Sacrificing Women: A Study of Ritualized Practices among Women Volunteers in India." *Voluntas: International Journal of Voluntary and Nonprofit Organizations* 19:242–267.

Mocellin, Jane S. P. and Peter Suedfeld. 1991. "Voices from the Ice: Diaries of Polar Explorers." *Environment and Behavior* 23:704–722.

Nansen, Fridtjof Dr. 1898. *Farthest North: Being the Record of a Voyage of Exploration of the Ship "Fram" 1893–96 and of a Fifteen Months' Sleigh Journey by Dr. Nansen and Lieut. Johansen* (Popular Edition). New York and London: Harper & Brothers Publishers.

Palinkas, Lawrence A. 2003. "The Psychology of Isolated and Confined Environments: Understanding Human Behavior in Antarctica." *American Psychologist* 58:353–363.

Palinkas, Lawrence C. and Peter Suedfeld. 2008. "Psychological Effects of Polar Expeditions." *The Lancet* 371:153–163.

Parry, Richard. 2001. *Trial by Ice: The True Story of Murder and Survival on the 1871 Polaris Expedition.* New York: Ballantine Books.

Peri, Antonio, Cristina Scarlata and Marta Barbarito. 2000. "Preliminary Studies on the Psychological Adjustment in the Italian Antarctic Summer Campaigns." *Environment and Behavior* 32:72–83.

Phillips, Bernard. 2001. *Beyond Sociology's Tower of Babel: Reconstructing the Scientific Method.* New York: Aldine de Gruyter.

_____. 2009. *Armageddon or Evolution? The Scientific Method and Escalating World Problems.* Boulder, CO: Paradigm Publishers.

Prior, Lindsay. 2004. "Following in Foucault's Footsteps: Text and Context in Qualitative Research." pp. 317–333 in *Approaches to Qualitative Research: A Reader on Theory and Practice,* edited by S. N. Hesse-Biber and P. Leavy. New York: Oxford.

Ronne, Finn. 1949. *Antarctic Conquest: The Story of the Ronne Expedition 1946–1948.* New York: G. P. Putnam's Sons.

Sarabia, Daniel and J. David Knottnerus. 2009. "Ecological Stress and Deritualization in East Asia: Ritual Practices During Dark Age Phases." *International Journal of Sociology and Anthropology* (Open Access Online Journal): 1(1):012–025, May. www.academicjournals.org/IJSA/contents/2009cont/May.htm

Scholes, Donald. 1951. *Fourteen Men.* London: George Allen & Unwin LTD.

Scott, Robert Falcon (edited by Max Jones). [1913] 2006. *Robert Falcon Scott Journals: Captain Scott's Last Expedition.* Oxford: Oxford University Press.

Sell, Jane, J. David Knottnerus, Christopher Ellison and Heather Mundt. 2000. "Reproducing Social Structure in Task Groups: The Role of Structural Ritualization." *Social Forces* 79:453–475.

Sell, Jane, J. David Knottnerus and Christina Adcock-Azbill. 2013. "Disruptions in Task Groups." *Social Science Quarterly* 94:715–731.

Sen, Basudhara and J. David Knottnerus. 2016. "Ritualized Ethnic Identity: Asian Indian Immigrants in the Southern Plains." *Sociological Spectrum* 36:37–56.

Shackleton, Sir Ernest. 1999 [1919]. *South: The Story of Shackleton's Last Expedition 1914–1917.* New York: Konecky & Konecky.

Simpson, C. J. W. 1957. *North Ice: The British North Greenland Expedition.* London: Hodder and Stoughton.

Strauss, Anslem L. 1987. *Qualitative Analysis for Social Scientists.* New York: Cambridge.

Stuster, Jack. 1996. *Bold Endeavors: Lessons from Polar and Space Exploration.* Annapolis, Maryland: Naval Institute Press.

Suedfeld, Peter and G. Daniel Steel. 2000. "The Environmental Psychology of Capsule Habitats." *Annual Review of Psychology* 51:227–253.

Suedfeld, Peter and Karine Weiss. 2000. "Natural Laboratory and Space Analogue for Psychological Research." *Environment and Behavior* 32:7–17.

Thornburg, P. Alex, J. David Knottnerus and Gary R. Webb. 2007. "Disaster and Deritualization: A Re-Interpretation of Findings from Early Disaster Research." *The Social Science Journal* 44:161–166.

_____. 2008. "Ritual and Disruption: Insights from Early Disaster Research." *International Journal of Sociological Research* 1:91–109.

Todd, A.L. 1961. *Abandoned: The Story of the Greely Arctic Expedition 1881–1884*. New York: McGraw Hill.

Ulsperger, Jason S. and J. David Knottnerus. 2007. "Long-term Care Workers and Bureaucracy: The Occupational Ritualization of Maltreatment in Nursing Homes and Recommended Policies." *Journal of Applied Social Science* 1:52–70.

_____. 2008. "The Social Dynamics of Elder Care: Rituals of Bureaucracy and Physical Neglect in Nursing Homes." *Sociological Spectrum* 28:357–388.

_____. 2009a. "Illusions of Affection: Bureaucracy and Emotional Neglect in Nursing Homes." *Humanity and Society* 33:238–259.

_____. 2009b. "Institutionalized Elder Abuse: Bureaucratic Ritualization and Transformation of Physical Neglect in Nursing Homes." pp. 134–155 in *Bureaucratic Culture and Escalating World Problems: Advancing the Sociological Imagination*, edited by J. David Knottnerus and Bernard Phillips. Boulder, CO: Paradigm Publishers.

_____. 2010. "Enron: Organizational Rituals as Deviance." pp. 291–294 in *Readings in Deviant Behavior*, 6th ed., edited by Alex Thio, Thomas C. Calhoun and Addrain Conyers. New York: Allyn and Bacon.

_____. 2011. *Elder Care Catastrophe: Rituals of Abuse in Nursing Homes – and What You Can Do About It*. Boulder, CO: Paradigm Publishers.

_____. 2013. "Care Giving Without the Care: The Deviant Treatment of Residents in Nursing Homes." pp. 209–219 in *Deviance Today*, 1st ed., edited by Alex Thio, Thomas C. Calhoun and Addrain Conyers. Upper Saddle River, NJ: Pearson.

_____. 2021. "Care Giving Without the Care: The Deviant Treatment of Residents in Nursing Homes" (New Chapter). pp. 377–390 in *Deviance Today*, 2nd ed., edited by Addrain Conyers and Thomas C. Calhoun. New York and London: Routledge.

Ulsperger, Jason S., J. David Knottnerus and Kristen Ulsperger. 2014. "Bureaucratic Rituals in Nursing Homes: The "CARE Model" and Culture Change." *The International Journal of Aging and Society* 3(3):21–33 (on-line).

_____. 2017. "The Importance of Rituals in Understanding Mass Homicide: A Structural Ritualization Analysis of the Ronald Gene Simmons Murders." pp. 199–230 in *Rituals: Past, Present and Future Perspectives*, edited by Edward Bailey. New York: Nova Science Publishers.

Van De Poel-Knottnerus, Frederique and J. David Knottnerus. 1994. "Social Life through Literature: A Suggested Strategy for Conducting a Literary Ethnography." *Sociological Focus* 27:67–80.

_____. 2002. *Literary Narratives on the Nineteenth and Early Twentieth-Century French Elite Educational System: Rituals and Total Institutions*. Lewiston, NY: Edwin Mellen Press.

_____. [2011] 2016. "Disruption and Deritualization: Concentration Camp Internment and the Breakdown of Social Order." pp. 107–131 in *Ritual as a Missing Link: Sociology, Structural Ritualization Theory and Research*, edited by J. David Knottnerus. New York and London: Routledge.

Varner, Monica K. and J. David Knottnerus. 2002. "Civility, Rituals and Exclusion: The Emergence of American Golf During the Late Nineteenth and Early Twentieth Centuries." *Sociological Inquiry* 72:426–441.

_____. 2010. *American Golf and the Development of Civility: Rituals of Etiquette in the World of Golf.* Koln, Germany: LAP Lambert Academic Publishing.

Walton, Kevin E. W. 1955. *Two Years in the Antarctic.* New York: Philosophical Library.

Warner, W. Lloyd. 1959. *The Living and the Dead: A Study of the Symbolic Life of Americans.* Yankee City Series. New Haven, CT: Yale University Press.

_____. 1962. *American Life: Dream and Reality.* Rev. ed. Chicago: University of Chicago Press.

Weiss, Karine, Peter Suedfeld, G. Daniel Steel and Masafumi Tanaka. 2000. "Psychological Adjustment During Three Japanese Antarctic Research Expeditions." *Environment and Behavior* 32:142–156.

Wilson, Robert N. 1986. *Experiencing Creativity: On the Social Psychology of Art.* New Brunswick, NJ: Transaction Books.

Wu, Yanhong and J. David Knottnerus. 2005. "Ritualized Daily Practices: A Study of Chinese 'Educated Youth'." *Shehui (Society)* (6):167–185 (Chinese academic journal).

_____. 2007. "The Origins of Ritualized Daily Practices: From Lei Feng's Diary to Educated Youth's Diaries." *Shehui (Society)* (1):98–119.

INDEX

Note: Italicised folios refers figures, bold tables and with "n" refers notes in the text.